Power Systems

More information about this series at http://www.springer.com/series/4622

Ping Ju

Stochastic Dynamics of Power Systems

Springer

Ping Ju
College of Energy and Electrical
Engineering
Hohai University
Nanjing, China

ISSN 1612-1287 ISSN 1860-4676 (electronic)
Power Systems
ISBN 978-981-13-1815-3 ISBN 978-981-13-1816-0 (eBook)
https://doi.org/10.1007/978-981-13-1816-0

Library of Congress Control Number: 2018949865

This Springer imprint is published by the registered company Springer Nature Singapore Pte Ltd.
The registered company address is: 152 Beach Road, #21-01/04 Gateway East, Singapore 189721, Singapore

Preface

Stochastic factors, widely existing in nature, engineering, and social systems, are also found in the power system. The renewable energy generation, electric vehicles, and power electronic devices are integrated into power grids at high proportion. As a result, the power system dynamics are facing increasingly significant influences of stochastic factors. In the seventies of the last century, the consideration of stochastic elements emerged in researches on the power system. However, these researches focused more on steady-state stochastic issues, leaving dynamic ones being definitive. As stochastic factors increasingly pile up, researches on dynamic issues under stochastic disturbances are becoming necessary.

Therefore, the writer wrote this book on the basis of study results in power system stochastic dynamics achieved in recent years. This book consists of seven chapters in total. Chapter 1 gives an introduction. Chapter 2 presents fundamental theories of stochastic dynamics. Chapters 3–7 set forth the model simulation, stability, dynamic oscillation, dynamic security, and optimal control of the power system stochastic dynamics.

Chapters 3–7 assemble the research results of our team. Great thanks to my team members, YU Yiping, ZHOU Haiqiang, SUN Lixia, WU Feng and WANG Chong et al., who contributed so much in researches and compilation of this book. They deserve to be entitled as authors of it. Meanwhile, many thanks also to graduate students LIU Yongfei, LI Hongyu, LIN Xue, ZHANG Jianyong et al., who participated in partial researches. I would like to express my special appreciation to Mr. ZHU Weiqiu, Academician of Chinese Academy of Science in Zhejiang University, for his kind guidance on relevant researches of our team, and his permission that we refer to the book *Introduction to Stochastic Dynamics* authored by him. Our researches were funded and supported by National Program on Key Basic Research Project (973 Program) Subject (2013CB228204), Key Program of National Natural Science Foundation of China (No. 51137002), Major Program Subject of National Natural Science Foundation of China (No. 51190102), Key Research Program Jiangsu Natural Science Foundation (No. BK2011026), etc. The author would also thank the "111" project of "Renewable Energy and Smart Grid" (B14022), and the State Grid Corporation of China together with its affiliate power grids, and provincial power grid corporations who provided us with application opportunities, especially Henan, Zhejiang and Jiangsu power grids. Researchers in this book also benefited from many experts and Hohai university. Here, I would like to express my true gratitude for all of you!

Please do not hesitate to let me know if any content in need of improvement attributed to the author's limited theory level and lack in practical experiences in this book. Feedbacks from all my readers will be highly appreciated! (e-mail: pju@hhu.edu.cn).

Nanjing, China
May 2018

Ping Ju

Contents

Chapter 1
Introduction

Abstract The stochastic dynamics of power system is a new field that requires a top-down design. In this chapter, a research framework is constructed for that purpose, based on the analysis of the background, transformations, and stochastic properties of power systems. Proposals for the research are provided by focusing on issues with the dynamics of power systems subjected to stochastic factors, including the stochastic dynamics model and the stochastic features of power systems (e.g., dynamic response, stability, security, and optimal control).

1.1 Background of the Stochastic Dynamics of Power System

Probabilistic and statistical methods are adopted in stochastic dynamics to study various stochastic dynamic processes and phenomena in nature, engineering and social systems [1, 2]. Stochastic dynamics originated from the quantitative description of the Brownian motion proposed by Einstein and others in the early twentieth century. During the 1940s and 1960s, the stochastic noise theory, stochastic vibration, and stochastic structural dynamics had been developed one after another. Since 1960, mathematicians have been developing the theory of stochastic stability and stochastic optimal control, which was used in economic and financial fields. Since 1980, stochastic dynamical system theories, such as the stochastic bifurcation theory, have been developed. One might say that by 1970 the stochastic dynamics theory of linear systems had grown to maturity. Stochastic dynamic theories of nonlinear systems have provided numerous methodologies, but many challenges still remain to be resolved.

Randomness always exists in power systems. However, because it had not yet become such an important factor in the operational security of power systems and the theoretical methods as well as the computational speeds were insufficient, traditional power system studies remained mostly deterministic. Among which, the study of deterministic power system dynamics has achieved considerable results,

P. Ju, *Stochastic Dynamics of Power Systems*, Power Systems,
https://doi.org/10.1007/978-981-13-1816-0_1

mainly including power system dynamic modeling [3, 4], power system dynamic analysis [5, 6], and power system control [7], etc.

Following the increasing integration of renewable energy and electric vehicles [8, 9], the stochastic issue will gradually become a prominent, communal and fundamental one for power dynamics in the future. But, the traditional power system dynamics theory, based on deterministic models, has obvious limitations, since it fails to consider the influence of stochastic factors on the dynamic characteristics of power systems. Therefore, it is crucial and urgent to carry out research in the stochastic dynamics of power systems.

In this chapter, the stochastic properties of power systems are briefly analyzed, and then a framework for stochastic dynamics research is proposed. Furthermore, it focuses on the stochastic dynamics model, stochastic dynamic response, stochastic stability, and stochastic security, etc.

1.2 Stochastic Properties of Power System

1.2.1 New Changes to Power System

In response to the challenges of energy supply, transmission, allocation, and development quality emerging in China's energy development, the State Grid Corporation of China has proposed the construction of a durable and smart power grid system, which is ubiquitously and globally interconnected, supported by a backbone of UHV power grids, and oriented toward the clean energy transmission [8]. Compared with traditional power grids, the smart grid brings new changes with regard to the power supply, electrical load, and network structure, etc., leading to new challenges to power system analysis [9]. The new changes of future power systems are summarized as "three high proportions, three multiples, and three characteristics" in this book.

"Three high proportions" refers to: (1) a high proportion of new energy power generation, which is predicted to reach 50% by 2030; (2) a high proportion of new loads, including electric vehicles and inverter air conditioners, etc.; (3) a high proportion of power electronic equipment, including power electronics used to connect the renewable energy to the grid, DC equipment in the transmission and distribution networks, FACTS technologies, and control tools based on power electronics in some new loads.

"Three multiples" refers to: (1) multiple interconnected energy sources, including electricity, gas, heat, and other energy sources; (2) multiple interconnected networks, including power networks, information networks, transportation networks, and meteorological networks, etc.; (3) multiple competitors, including diversified suppliers on the generation side and various choices on demand side, etc.

"Three characteristics" refers to: (1) intelligence, meaning that the power grid will gradually move toward automaticity; (2) flexibility, meaning that the power grid can deal with various disturbances so as to maintain continuous power supply; (3) stochastic nature, meaning a variety of stochastic phenomena occurring in the power grid.

1.2.2 Stochastic Properties of Power System

Stochastic properties are those that possess the quality of uncertainty, which follows statistical rules. Stochastic factors are uncertain and follow statistical rules, including stochastic disturbances, stochastic initial states, and stochastic parameters, etc. Stochastic disturbances often refer to relatively fast-varying stochastic factors, including the stochastic disturbance and stochastic faults, etc.

Regarding the origin of stochastic factors, there are three major aspects: (1) renewable energy generation, such as the wind power and solar power generation, etc.; (2) network events, such as the network operation and network fault, etc.; (3) load variations, such as electric vehicles and electric trains, etc.

In terms of types of stochastic factors, there are two types: (1) Continuous stochastic disturbances, which consist of load changes and renewable energy generation changes, etc. (2) Discrete stochastic events, which include fault types and locations, network topologies, and power electronic actions, etc.

As for the time scale, several stochastic factors can be presented in the following formula:

$$\begin{cases} \frac{\mathrm{d}X(t)}{\mathrm{d}t} = \boldsymbol{F}_x[\boldsymbol{X}(t), \boldsymbol{U}(t), \boldsymbol{\theta}, t] + \boldsymbol{G}_x[\boldsymbol{X}(t), \boldsymbol{\theta}, t]\boldsymbol{\varepsilon}(t) \\ \boldsymbol{Y}(t) = \boldsymbol{F}_y[\boldsymbol{X}(t), \boldsymbol{\theta}, t] + \boldsymbol{G}_y[\boldsymbol{X}(t), \theta, t]\boldsymbol{\varepsilon}(t) \\ \boldsymbol{X}(0) = \boldsymbol{X}_0 \end{cases} \tag{1.2.1}$$

where $\boldsymbol{X}$ represents a status vector; $\boldsymbol{U}$ is an input vector; $\boldsymbol{Y}$ is an output vector; $\boldsymbol{\theta}$ is a parameter vector; $\boldsymbol{F}$ and $\boldsymbol{G}$ are non-linear function vectors; $\boldsymbol{X}_0$ means an initial value vector; and $\boldsymbol{\varepsilon}(t)$ is an external stochastic disturbance vector.

In the above formula, stochastic properties can be classified into 3 aspects:

(1) The stochastic factor of external disturbance $\boldsymbol{\varepsilon}$: a stochastic factor which varies quickly on a timescale of seconds, mainly originating from the continuous stochastic power fluctuation during power system operations. It is described by way of a dynamic stochastic process. The characteristic of external stochastic disturbance is rapidly changing, which may induce the dynamic process in the power system. For example, a continuous stochastic power fluctuation from the renewable energy generation or electric vehicles can cause stochastic fluctuations in the electromagnetic and mechanical power components of the rotary

motion equation, and thus a power imbalance between the two, which is in turn presented as an additive stochastic process in the rotary motion equation.

(2) The stochastic factor of parameters $\boldsymbol{\theta}$: a stochastic factor which varies slowly on a timescale of minutes, mainly originating from parameters for those stochastic factors acting on the system model, such as the model error, state estimation error, short-term load forecasting error and the intermittency of wind power. Among the above mentioned stochastic parameters, the generator damping coefficient and load characteristic coefficient, etc., have high-degree randomness.

(3) The stochastic factor of initial value $\boldsymbol{X}_0$: On one hand, a quasi-steady-state stochastic factor, is mainly from slow changes in power generation. On the other hand, it also indicates that the system has a stochastic initial state after a stochastic disturbance. If a line is confirmed to be tripped while a fault occurs, because the fault clearing time is stochastic, the system status would be stochastic as a result when the fault is cleared. This state can be regarded as the initial value of the subsequent autonomous process.

1.3 Research Framework of Stochastic Dynamics of Power System

1.3.1 Research Status and Framework

Power system steady-state problems can be described by a set of algebraic equations. The stochastic property of the generation and the load has been considered in steady-state problems, for example, with respect to the power system planning, flow, and optimization, etc.

Issues of power system dynamics can be presented by a set of differential algebraic equations. It should be mentioned that there are plenty of valuable research results under the deterministic framework in power system dynamics. Accordingly, the analysis of power system probabilistic stability, taking the fault probability into account, has been researched well, which means the system stability probability is calculated based on the fault probability [6, 10, 11]. Recently, the analysis of the power system dynamic response under stochastic disturbance has attracted much attention. Monte Carlo simulation based on the numerical method provides an acceptable tool [12–14], which is widely applicable, but the calculations are enormous and the mechanism is unclear. Therefore, analytical methods with the less calculation and clearer mechanism have been studied [15, 16].

To sum up, the study of dynamic issues in power systems under stochastic disturbances is at the beginning. Are there any new dynamic phenomena in the

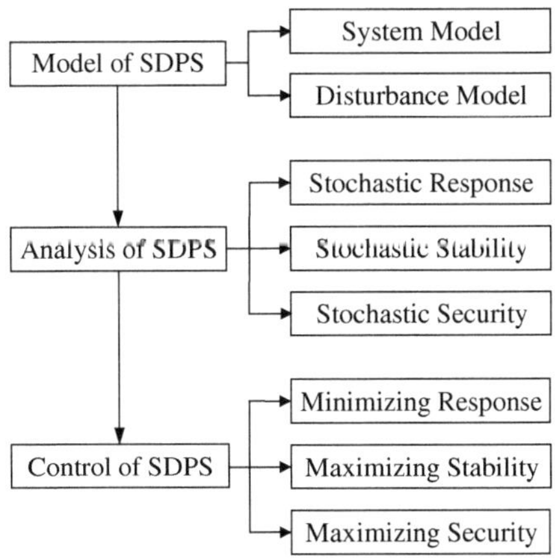

Fig. 1.1 The research framework of stochastic dynamics of power system

power system under stochastic disturbance? Will new security problems happen? What specific tools can be taken to suppress the stochastic disturbances? All above questions require being further researched.

The Stochastic Dynamics of Power System (SDPS) is a power system's dynamic behavior under the influence of stochastic factors. Three major research areas are shown in Fig. 1.1.

(1) Models of SDPS. Studies of stochastic dynamic models which take stochastic factors into account must focus not only on how to include stochastic factors in the system models but also on how to describe the stochastic disturbance itself.
(2) Analysis of SDPS. Firstly, it is necessary to analyze the stochastic dynamic response of the power system, which involves response characteristics and the calculation of the power system under a stochastic disturbance. Then, the stochastic stability of the power system should be analyzed, including the probability or statistical stability of the power system under the stochastic disturbance. Finally, the stochastic dynamic security of the power system should be analyzed, which relates to the safe operation of the system under the stochastic disturbance.
(3) Control of SDPS. Stochastic dynamic control indicates that [17] the controller, according to the latest information of the system state, selects an optimal control, from among all the possibilities which satisfy the constraint parameters, and thereby generates the ideal outcome for the controlled system's predetermined target. Indexes of stochastic dynamics can be analyzed according to the predetermined targets, such as the minimized response, the maximized stability, the maximized dynamic security, etc. Furthermore, stochastic control research should be carried out in terms of actual situations such as partial observations, uncertainty models, and time lag, etc.

Table 1.1 Comparative study of dynamics of power system

Dynamics	Deterministic dynamics	Stochastic dynamics
Target	Dynamic characteristics of a power system under a specific disturbance	Dynamic characteristics of a power system under stochastic disturbance
Equation	Deterministic differential algebraic equations	Stochastic differential algebraic equations
Methods	Numerical simulation or direct method	Monte Carlo simulation or analytical method

The above three aspects are interconnected and sequential. The stochastic dynamic models provide a foundation for the latter two aspects. On this basis, stochastic dynamical characteristics of the power system are analyzed and in turn improved through effective controls.

Table 1.1 shows a comparison between the stochastic dynamics and deterministic dynamics of a power system. It should be noted that the stochastic dynamics is not a substitute for, but a supplement to the deterministic dynamics of the power system. If the stochastic disturbance is not significant, the deterministic dynamics will suffice. Moreover, the stochastic dynamics of the power system is based on its deterministic dynamics. For instance, although a stability quantification theory [5] and a "domain" methodology [6] are proposed for a deterministic power system, it can also be applied to the study of stochastic dynamics of power systems [18].

1.3.2 *Study of Stochastic Dynamic Model of Power System*

1. **System Model**

Slow-varying stochastic factors can be described through stochastic algebraic equations (SAE) for the power system. Such research is plentiful, including stochastic planning, stochastic flow, and stochastic optimizing [19–21]. Fast-varying stochastic factors can be described by stochastic differential equations (SDE) as well as stochastic differential algebraic equations (SDAE). Corresponding research is in the initial stages and can draw on the theory of stochastic dynamics and other disciplines. Markov chains can be used to describe discrete stochastic factors [22, 23]. Markov process theory is a good choice for this immature study of discrete stochastic factors.

Currently, a time-domain model, which can be easily realized using computer programs, is adopted in most power system computation and analysis. However, time-domain curves obtained by the time-domain model are complicated for the stochastic properties of power systems. By contrast, frequency-domain curves obtained by using a frequency domain model are capable of describing the rules and characteristics of stochastic properties more clearly. Therefore, in the study of the stochastic dynamics of power systems, the time-domain model and frequency-domain model are of equal importance and are often used interchangeably. Naturally, it should be mentioned that the frequency-domain model is only applicable to a linear system, meaning the power system is subject to a small disturbance.

2. **Stochastic Disturbance Model**

After the power system is modeled, stochastic disturbance comes next, which must be modeled properly based on the disturbance characteristics [24]. The stochastic disturbance in the power system can be described as a stochastic process, which may be regarded as a series of related stochastic variables at different moments. The stochastic process can be classified in terms of different characteristics [25]. Regarding its probability distribution, the stochastic process can be classified as a Gaussian process, a Rayleigh process, and a Poisson process, etc. In terms of the bandwidth of its spectral density, the stochastic process includes a narrow band process, a broadband process, and white noise of an infinite bandwidth. It consists of a stationary process and a non-stationary process according to its evolution in time. In power system research, it is currently assumed that the stochastic process is a Gaussian distribution white noise [26]. The assumption is for convenience in analysis on one hand, and for significant development of the theory based on ideal distribution on the other hand, which provides researchers with a series of analytical tools [15]. However, actual random characteristics are expected to be utilized, because the actual stochastic process is often not an ideal Gaussian distribution or white noise. This requires statistical analysis of years of data accumulation, or a reasonable simplified assumption based on reality. In addition to Monte Carlo simulation, we also need to research the method of analyzing a stochastic disturbance that is not a Gaussian distribution or white noise.

3. **Model Conversion**

The above model sometimes is difficult to use directly in research. Therefore, in order to exploit some superior theoretical tools such as a stochastic averaging method, the above model needs to be converted to a corresponding model, such as a dissipative quasi-Hamiltonian system model under a stochastic disturbance [2]:

$$\begin{cases} \frac{\mathrm{d}Q_i}{\mathrm{d}t} = \frac{\partial H}{\partial P_i} \\ \frac{\mathrm{d}P_i}{\mathrm{d}t} = \frac{\partial H}{\partial Q_i} - c_{ij}(\boldsymbol{Q},\boldsymbol{P})\frac{\partial H}{\partial P_j} + f_{ik}(\boldsymbol{Q},\boldsymbol{P})\xi_k(t) + u_i(\boldsymbol{Q},\boldsymbol{P}) \\ i,j = 1,2,\ldots n;\ k = 1,2,\ldots,m \end{cases} \tag{1.3.1}$$

where $\boldsymbol{Q}$ and $\boldsymbol{P}$ are a generalized displacement vector and a generalized velocity vector respectively, u is a control variable, H is a Hamiltonian function, ξ is a stochastic disturbance, and c_{ij} and f_{ik} are coefficients.

1.3.3 Study of Stochastic Dynamic Response of Power System

1. **Concept of Stochastic Dynamic Response of Power System**

In deterministic problems, responses of a dynamic system consist of a transient response and a steady-state response. The transient response is a system response right after applying a disturbance, where an initial condition plays an important role. The steady-state response is a system response after applying a long enough disturbance, where an initial condition has no effect. Corresponding to the steady-state response in a deterministic system, it is called a stationary response in a stochastic system. In a given system, there may or may not be a stationary response, depending on whether the disturbance is a stationary process or a non-stationary process. Since the non-stationary stochastic process is difficult to deal with, the stationary response under a steady stochastic disturbance is usually studied.

It should be noted that "static state" strictly means that a variable is constant; and "steady state" refers to a final steady state of the system, which can be static or steadily changes (such as a limit cycle); "stationary response" refers to a final response under a steady stochastic disturbance.

The traditional dynamic characteristics of a power system are characteristics represented by the system's dynamic response subject to a given disturbance (such as the relative rotor angle of a generator, or tie-line power, etc.) [27]. The oscillation problems matter the most. Under a stochastic disturbance, the computation and the oscillation problems of the dynamic response in the power system are to be studied accordingly.

2. **Computation of Power System Dynamics under Stochastic Disturbance**

At present, the study of the dynamic process computation method in power systems under stochastic disturbance can be divided into two situations. Firstly, for a small-scale power system, a common approach is to approximate the stochastic disturbance as a Gaussian white noise process, introduce stochastic items directly into a differential equation model of the power system, and then describe the system dynamic response under the stochastic disturbance as the solution to an SDE [28]. Taking account of discrete stochastic events, such as system failures, the stochastic disturbance can be assumed to be a Poisson stochastic process and solved in combination with the above SDE model [29]. At present, relative mature SDE numerical solutions include the Euler-Maruyama, Milstein, Heun, and Runge-Kutta methods, among others [30]. Secondly, for a large-scale power system, due to its numerous components and complex structures, the SDAE model is of such a high order that it is difficult to solve numerically. A currently common large-scale power grid model is a time-domain model established via simulation platforms such as PSASP and Matlab. The stochastic disturbance can be described as a set of stochastic time sequences, and then injected into a corresponding node in the system by a certain method, such as a time-varying current injection method [31], for inclusion in the simulating calculation.

Convergence is an important indicator for evaluating a numerical calculation method [30]. There are two convergence indicators when solving for SDE values. One indicator is that the trajectory of the numerical solution is required to be as close as possible to the real trajectory when a research problem involves the numerical simulation of the solution process. In this case, the strong convergence of the numerical method should be the point of focus. The other one is that, if the study does not target the trajectory of the solution process, but only momentary characteristics of the solution process, the weak convergence of the numerical method is the point of focus. Generally, a numerical method with good convergence is relatively complicated in terms of calculation steps, requiring many more calculations and being more time-consuming. However, a numerical solution with simple calculations, usually simplified by certain assumptions, is weak in convergence. In comparison to the commonly used Euler-Maruyama method for calculating the simple stochastic dynamic response of a power system, the Heun method is proven to have a higher error convergence order and better numerical stability [32].

3. **Power System Oscillation under Stochastic Disturbance**

Current explanations of power system oscillation mechanisms mainly include the negative damping mechanism, forced oscillation mechanism, mode resonance mechanism, etc. [33]. But most of these mechanism explanations are based on a deterministic differential equation model and cannot be used to analyze the effect of stochastic disturbances on the system's dynamic response. Consequently, traditional time-domain methods have limitations with regard to the mechanism of power system oscillation under stochastic disturbance.

For power system oscillation under a small stochastic disturbance, the linearized system can be analyzed from the perspective of the frequency domain. The author has found the phenomenon of general forced oscillation and its mechanism [34], pointing out that if the power spectrum band of stochastic disturbance covers the natural oscillation frequency corresponding to some weak damping oscillation modes of the system, a larger oscillation would be induced. The induced oscillation is also a narrow-band stochastic process, including these weak damping mode frequency components. The traditional forced oscillation conditions can be met if a disturbance frequency is exactly equal to or approximately equal to a natural oscillation frequency of the system when the disturbance source has a single frequency. This "point-to-point" oscillatory condition indicates that the traditional forced oscillation is of small probability. Those induced by stochastic disturbances belong to forced oscillation, which is a kind of "face-to-point" oscillation condition that can only be met when some weak damping mode frequencies are covered by the stochastic disturbance frequency band. This significantly increases the possibility of the forced oscillation. Therefore, the forced oscillation induced by the stochastic disturbance is called a "general forced oscillation" while the traditional forced oscillation may be regarded as an exception called a "special forced oscillation."

The nonlinear system theory may be applied to the analysis of the power system's oscillation under a large stochastic disturbance, because the system's operating state will deviate away from its stable equilibrium point as the disturbance grows stronger, presenting stronger nonlinear characteristics. When the nonlinear characterization of the power system is strong, the oscillation modes of the system fail to be completely decoupled but have a certain nonlinear coupling interaction. Research has shown that in a deterministic nonlinear system with multiple degrees of freedom, internal resonance will occur due to the effect of proper nonlinear coupling when one of the system's natural frequencies becomes an integer multiple of another [35]. When a weak damping oscillation happens in the power system, a 1:2 internal resonance may be induced in the system by nonlinear coupling between oscillation modes [36]. At this moment, real dangerous modes may be hidden, resulting in inappropriate control strategies. The internal resonance phenomenon is also present in the forced oscillation phenomenon of the power system caused by the stochastic disturbance. However, further study is necessary due to its highly complex mechanism.

1.3.4 Study of Stochastic Stability of Power System

1. **The Concept of Stochastic Stability of Power System**

The traditional power system stability refers to a power system's ability to return to its original steady-state operation or reach a new one after being disturbed based on the inherent system capability and the effect of device control [37].

The traditional power system stability analysis is carried out in the case of given component parameters, operating status, and disturbance modes, which is part of the deterministic stability analysis. In the probabilistic stability analysis of power systems, the stability probability index of the system is determined through the statistical characteristics of the main stochastic factors that affect the stability, which is a crucial supplement to the traditional stability analysis. However, the above study only takes into account the stochastic property caused by parameter changes and faults in the power grid, ignoring the stochastic factors such as stochastic disturbance.

The dynamic response of the power system under a stochastic disturbance is also a stochastic process, so the stochastic stability condition will be given in the form of probability statistics. The power system model considering the stochastic disturbance is rendered as an SDE. The concept of SDE stability can draw in [25] Lyapunov stability of probability 1, Lyapunov asymptotic stability of probability 1, the probability stability, probability asymptotic stability, M-moment stability, and M-moment asymptotic stability.

2. **Power System Stability under Small Stochastic Disturbance**

Under a small stochastic disturbance, the power system can approximate a linearized model.

The mean value stability is a first-moment stability problem. Provided the power system under a stochastic disturbance has mean value stability, it becomes evident that under the stochastic disturbance, the mean value of a power system dynamic response will remain within a small neighborhood and interrupt any divergence which might cause system instability. Research has shown that [38, 39] as long as the power system has small disturbance stability (i.e., it has negative damping), the mean value of the power system will be stable under a Gaussian small stochastic disturbance.

It is not enough to consider only the mean value stability during analysis of the power system's stochastic stability. It is also necessary that the variance of the response process be bounded, which is a second-moment stability problem. The research shows that [38] once the power system has small disturbance stability, the mean value of the power system is stable under the Gaussian small stochastic disturbance.

Under the small stochastic disturbance, the mean value stability mainly indicates whether the mean value of the dynamic response is within the neighborhood of the steady-state value. The smaller the neighborhood's radius, the closer the dynamic mean value will be to the steady-state value. In addition, the mean square stability is used to describe the degree that the dynamic response deviates from the dynamic mean value, which is not directly related to the steady-state value. That is to say, the explanation of the mean square stability effect requires establishing a connection between the mean value stability and the steady-state value. The mean value stability is a precondition of the mean square stability. Thus, in order to effectively depict the stochastic stability characteristics of the system, the mean value stability and the mean square stability should be analyzed simultaneously.

3. **Power System Stability under Large Stochastic Disturbance**

Studies on the mean value stability and mean square stability are all based on the power system linearized stochastic model, and the combination of the small disturbance stability analysis and the stochastic system's moment stability analysis.

However, the linearized stochastic model is not applicable to stochastic disturbances of great intensity, such as stochastic faults together with stochastic disturbances. Hence, a nonlinear stochastic model of the power system should be adopted. Few universal methods are useful for analyzing the nonlinear model's stochastic stability. But we can find corresponding solutions in combination with specific problems, such as simplifying the system via the extended equal area criterion and other direct methods, or establishing its low-dimensional stochastic differential equation model, then using analytic methods for analysis.

4. **Stochastic Bifurcation of Power System**

The bifurcation theory studies a system's dynamic behavior determination or its topological changes when the system's characteristics change smoothly to pass through a critical threshold. Over the past few decades, the bifurcation theory has extended from deterministic systems to stochastic systems [25]. Similar to a deterministic bifurcation, a stochastic bifurcation is a phenomenon of sudden qualitative change in the behavior of a stochastic disturbance system when a slight smooth change occurs in system parameters. Taking the steady state into account is sufficient for a deterministic system. Regarding a two-dimensional system, its qualitative behaviors are characterized by equilibrium points (fixed points), stability or instability, periodic orbits, and limit cycles. In contrast to the steady state of the deterministic system, taking stability into account is sufficient for a stochastic system. Periodic orbits and limit cycles in the deterministic system no longer emerge in the stochastic system. Instead, the qualitative behavior of the system is analyzed in terms of so-called invariant measures, including steady or unstable fixed points and the stationary probability density. In fact, the fixed point corresponds to the probability density of δ. Thus, in this case, the invariant measure is equivalent to the stationary probability density, such that the qualitative change of the stationary probability density may be used to define a stochastic bifurcation phenomenon.

The deterministic bifurcation theory of the power system has already achieved a series of research outcomes [6]. But stochastic bifurcation problems of the power system under the stochastic disturbance need further and in-depth study.

1.3.5 Study of Stochastic Dynamic Security of Power System

1. **The Concept of Stochastic Dynamic Security of Power System**

Traditional power system security refers to the ability of a power system to withstand sudden disturbances [37]. Security, also known as dynamic reliability, is the capability of a power system to withstand sudden disturbances and to continually provide users with electricity and electric energy under dynamic conditions.

In the above traditional definitions of security, there are two vital points: "sudden disturbances", such as a sudden short circuit or unexpected trip of system components; and "dynamic conditions", which means differential equations are required to describe the system.

Taking the stochastic properties of a power system into account, the definition of security is in need of expansion. First of all, the disturbance includes not only "sudden disturbances" such as a sudden short circuit, but also continuous and sustaining disturbances such as a stochastic disturbance. When considering stochastic events, stochastic disturbances, and stochastic parameters, they can be divided into 1-dimensional problems (e.g., stochastic disturbances), 2-dimensional problems (e.g., stochastic events and stochastic disturbances), and 3-dimensional problems (e.g., stochastic events, stochastic disturbances and stochastic parameters). Additionally, security is no longer a deterministic problem, but a stochastic one. The indicators describing security are also no longer deterministic ones (for example, a certain power system is definitely safe or stable), but probabilistic ones (such as the security probability of a power system [40]). In summary, the definition of a power system's stochastic security refers to the capability of a power system to bear stochastic disturbances. The stochastic disturbances mentioned here involve traditional sudden disturbances as well as continuous stochastic disturbances.

2. **Probability within Dynamic Security Region**

Security, in essence, requires that the system operating status and certain indicators are maintained within a certain range. That is, the system operation is limited to a permissible region, such as maintaining a normal level of grid frequency and voltage.

Stochastic disturbance constantly exists in the power system, but under normal circumstances, its strength is insufficient to destabilize the system [32]. Therefore, investigating a system's security under stochastic disturbances is mainly about whether the system is constrained within a limited region Ω_B which is acceptable to its engineering, which may be called a "permissible region" or a "bounded fluctuation region" [26].

Because of the stochastic disturbance, the system state $\boldsymbol{Y}(t)$ at time t becomes a stochastic vector. Therefore, it is a stochastic issue whether $\boldsymbol{Y}(t)$ can be kept within a certain region. It needs to be described in terms of probability, that is, the intra-region probability. The intra-region probability is essentially the mathematical reliability, which means that a process's initial value was located within a certain region Ω and kept its probability within Ω during the interval $(0, t]$ [15]. The above problem is equivalent to a "first passage" one in stochastic dynamics [2, 25].

If the intra-region probability problem is directly analyzed on the basis of SDAE, a "curse of dimensionality" takes place, resulting in failure of the practical application. Therefore, it is possible to do research based on the system's energy function. When the system energy is less than an allowable given value (called an "energy bound"), the system state is regarded as within the allowable region. The system state is a rather high-dimension vector, whereas the energy is a one-dimensional scalar. The stochastic averaging method for the quasi-Hamiltonian system provides a fast and effective scheme to solve the intra-region probability of the power system [26, 42]. Although the research based on an "energy bound" is somewhat conservative, it greatly simplifies the problem. Besides, conservatism is not bad for the power system's security.

1.3.6 Study of Stochastic Dynamic Control of Power System

1. **The Concept of Stochastic Dynamic Control of Power System**

Subject to stochastic factors, the control problem of a dynamic system is mostly constructed on the basis of Lyapunov stability, for which proper Lyapunov functions are required, and stochastic stability criteria are utilized to judge the system's level of stability, thereby obtaining corresponding control strategy. Such controls act on a certain system state, not on probability indicators (such as power system reliability [16, 41] and the probability within stochastic dynamic security region [26, 42]). Stochastic dynamic control of a power system refers to a stochastic optimal control problem, which is to choose a control within the allowable control set so as to optimize its performance indicators. Stochastic dynamic analysis of the power system is the foundation of its stochastic control. The power system controls are often subject to some kind of constraint. The control targets can be expressed as a minimum or maximum value of a certain performance indicator, which mainly refers to the minimized response, maximized stability and maximized dynamic security. The status and control of the controlled system are all stochastic processes, and the performance indicator is usually the mathematical expectation or probability density of a given function. In conclusion, the control of power system stochastic dynamics depends on the probability distribution of the involved process rather than its trajectory.

2. **Stochastic Dynamic Control Strategy of Power System**

An optimal control strategy for the control of a power system under a stochastic disturbance may be studied using a combination of the quasi-Hamiltonian system stochastic averaging method and stochastic dynamic planning. The methodology is as follows. First, the stochastic dynamical model is transformed into a controlled quasi-Hamiltonian system model. Then, the model is transformed into an Itô stochastic differential equation for average controlled parts based on the stochastic averaging method. Next, determine the optimal control rules for predetermined targets, indicators determining performance, dynamic planning equations and constraints by using dynamic planning and the maximum principle. Finally, the two criteria, control efficacy and control efficiency, are introduced to evaluate the control strategy.

The research on the nonlinear stochastic optimal control of a quasi-Hamiltonian system provides a mathematical theory foundation for the stochastic dynamic control of the power system. However, it is important to consider the feasibility of the control strategy and the operability of the actual actuator for the stochastic dynamic control of the power system. The control force must not exceed the maximum actuator capability. Currently, studies have been carried out on feedback control with the maximized reliability of single-machine infinite bus system under the stochastic disturbance [16]. In any case, this is just a beginning. Further study of stochastic dynamic control of multi-machine power system is still necessary.

In general, the research on stochastic dynamics of power system is of theoretical significance. The research focused mainly on stochastic dynamic strategies for modeling, analyzing, and controlling power systems. The stochastic dynamic analyses mentioned mainly consist of response, stability and security analyses. Drawing on achievements in related fields (such as stochastic dynamics) is necessary for research into stochastic dynamics of power systems, as long as we combine classical concepts together with the actual needs of the power system.

References

1. Zhu, Weiqiu. 1992. *Random Vibration [M]*. Beijing: Science Press. (in Chinese).
2. Zhu, Weiqiu. 2003. *Hamiltonian Theoretical Framework of Nonlinear Stochastic Dynamics and Control [M]*. Beijing: Science Press. (in Chinese).
3. Ju, Ping. 2010. *The Theory and Method of Power System Modeling [M]*. Beijing: Science Press. (in Chinese).
4. Kundur, P. 1994. *Power System Stability and Control [M]*. New York, USA: McGraw-Hill Inc.
5. Xue, Yusheng. 1999. *Quantitative Theory for Motion Stability [M]*. Nanjing: Jiangsu Science and Technology Press. (in Chinese).
6. Yu, Yixin, and Chengshan Wang. 1999. *Theory and Method of Power System Stability [M]*. Beijing: Science Press. (in Chinese).
7. Lu, Qiang, Yuanzhang Sun, and Shengwei Mei. 2001. *Nonlinear Control Systems and Power System Dynamics [M]*. Boston: Kluwer Academic Publishers.
8. Research Group of Strategy of Energy Field of China. 2009. *Energy Technologies Roadmap to 2050 of China [M]*. Beijing: Science Press. (in Chinese).
9. Zhou, Xiaoxin, Zongxiang Lu, Yingmei Liu, et al. 2014. Development models and key technologies of future grid in China [J]. *Proceedings of the CSEE* 34 (29): 4999–5008. (in Chinese).
10. Ju, Ping, and Daqiang Ma. 1990. Probabilistic analysis of power system stability [J]. *Automation of Electric Power Systems* 14 (3): 18–23. (in Chinese).
11. Xue, Yusheng, Yongjun Wu, Yunyun Xie, et al. 2016. Power system stability analysis for intercurrent natural disasters [J]. *Automation of Electric Power Systems* 40 (4): 10–18. (in Chinese).
12. Li, G., H. Yue, M. Zhou, et al. 2014. Probabilistic assessment of oscillatory stability margin of power systems incorporating wind farms [J]. *International Journal of Electrical Power and Energy Systems* 58: 47–56.
13. Yuan, B., M. Zhou, G. Li, et al. 2015. Stochastic small-signal stability of power systems with wind power generation [J]. *IEEE Transactions on Power Systems* 30 (4): 1680–1689.
14. Bie, Zhaohong, and Xifan Wang. 1997. The application of Monte Carlo method to reliability evaluation of power systems [J]. *Automation of Electric Power Systems* 21 (6): 68–75. (in Chinese).
15. Zhu, W.Q. 2006. Nonlinear stochastic dynamics and control in Hamiltonian formulation [J]. *Applied Mechanics Reviews* 59 (4): 230–248.
16. Chen, Lincong, and Weiqiu Zhu. 2010. Feedback maximization of reliability of a simple power system under random perturbations [J]. *Journal of Dynamics and Control* 8 (1): 19–23. (in Chinese).
17. Zhu, Weiqiu, and Zuguang Ying. 2013. Advances in research on nonlinear stochastic optimal control of quasi-Hamiltonian systems [J]. *Advances in Mechanics* 43 (1): 39–55.
18. Qin, Chao, and Yixin Yu. 2014. Security region based probabilistic small signal stability analysis for power systems with wind power integration [J]. *Automation of Electric Power Systems* 38 (10): 43–48. (in Chinese).
19. Huang, Yang, Wei Hu, Yong Min, et al. 2014. Risk-constrained coordinative dispatching for large-scale wind-storage system [J]. *Automation of Electric Power Systems* 38 (9): 41–47. (in Chinese).
20. Li, Yining, Hao Wu, Huanhai Xin, et al. 2015. Power system probabilistic load flow based on generalized polynomial chaos methods [J]. *Automation of Electric Power Systems* 39 (7): 14–20. (in Chinese).

21. Li, Feng, Lizi Zhang, Jun Shu, et al. 2014. A flexible stochastic optimal scheduling approach considering intermittent power and energy storage [J]. *Automation of Electric Power Systems* 38 (5): 1–7. (in Chinese).
22. Hou, Kai, Yuan Zeng, Hongjie Jia, et al. 2015. Risk assessment and optimization method for Markov chain-based scheduling operations [J]. *Automation of Electric Power Systems* 39 (19): 142–148. (in Chinese).
23. Jiang, Le, Junyong Liu, Zhenbo Wei, et al. 2015. Running state and its risk evaluation of transmission line based on Markov chain model [J]. *Automation of Electric Power Systems* 39 (13): 51–57. (in Chinese).
24. Zhang, Xi, Chongqing Kang, Ning Zhang, et al. 2014. Analysis of mid/long term random characteristics of photovoltaic power generation [J]. *Automation of Electric Power Systems* 38 (6): 6–13. (in Chinese).
25. Zhu, Weiqiu, and Guoqiang Cai. 2015. *Introduction to Stochastic Dynamics [M]*. Beijing: Science Press. (in Chinese).
26. Li, Hongyu, Ping Ju, Yiping Yu, et al. 2015. Bounded fluctuations regions and analytic method of intra-region probability in power system under stochastic excitations [J]. *Proceedings of the CSEE* 35 (14): 3561–3568. (in Chinese).
27. Kundur, P., J. Paserba, V. Ajjarapu, et al. 2004. Definition and classification of power system stability—IEEE/CIGRE joint task force on stability terms and definitions [J]. *IEEE Transactions on Power Systems* 19 (2): 1387–1401.
28. Wang, K., and M.L. Crow. 2013. The Fokker-Planck equation for power system stability probability density function evolution [J]. *IEEE Transactions on Power Systems* 28 (3): 2994–3001.
29. Dong, Z.Y., J.H. Zhao, and D. Hill. 2012. Numerical simulation for stochastic transient stability assessment [J]. *IEEE Transactions on Power Systems* 27 (4): 1741–1749.
30. Higham, D.J. 2001. An algorithmic introduction to numerical simulation of stochastic differential equations [J]. *Society for Industrial and Applied Mathematics* 43 (3): 525–546.
31. Chen, Daxuan, Yiping Yu, Ping Ju, et al. 2014. Modeling of special power load based on time-varying current injection method [J]. *Electric Power Automation Equipment* 34 (3): 120–124. (in Chinese).
32. Liu, Yongfei, Ping Ju, Yusheng Xue, et al. 2014. Calculation analysis on power system characteristics under random excitation [J]. *Automation of Electric Power Systems* 38 (9): 137–142. (in Chinese).
33. Klein, M., G.J. Rogers, and P. Kundur. 1991. A fundamental study of inter-area oscillations in power systems [J]. *IEEE Transactions on Power Systems* 6 (3): 914–921.
34. Ju, Ping, Yongfei Liu, Hongyin Wang, et al. 2014. General forced oscillations of power systems [J]. *Electric Power Automation Equipment* 34 (5): 1–6. (in Chinese).
35. Sanchez-Gasca, J.J., V. Vittal, M.J. Gibbard, et al. 2005. Inclusion of higher order terms for small-signal (modal) analysis: committee report-task force on assessing the need to include higher order terms for small-signal (modal) analysis [J]. *IEEE Transactions on Power Systems* 20 (4): 1886–1904.
36. Chen, Lei, Yong Min, and Yiping Chen. 2014. Study on internal resonance in power system low frequency oscillation [J]. *Power System Technology* 38 (1): 160–167. (in Chinese).
37. Editorial Committee. 2014. *China Electric Power Encyclopedia [M]*. Beijing: China Electric Power Press. (in Chinese).
38. Zhang, J.Y., P. Ju, Y.P. Yu, et al. 2012. Responses and stability of power system under small Gauss type random excitation [J]. *Science China Technological Sciences* 55 (7): 1873–1880.
39. Zhou, Ming, Bo Yuan, Xiaoping Zhang, et al. 2014. Stochastic small signal stability analysis of wind power integrated power systems based on stochastic differential equations [J]. *Proceedings of the CSEE* 34 (10): 1575–1582. (in Chinese).
40. Liu, Yanli, and Yixin Yu. 2013. Probabilistic steady-state and dynamic security assessment of power transmission system [J]. *Science China Technological Sciences* 43 (5): 523–532.

41. Zhao, Yuan, Jie Wang, Yanjiao Xiong, et al. 2016. A Gaussian mixture model of multivariable in power system reliability assessment [J]. *Automation of Electric Power Systems* 40 (1): 66–71 + 80. (in Chinese).
42. Li, Hongyu, Ping Ju, Xinqi Chen, et al. 2015. Stochastic averaging method for quasi Hamiltonian system of multi-machine power systems [J]. *Science China Technological Sciences* 45 (7): 766–772.

Chapter 2
Fundamentals of Stochastic Dynamics

Abstract After decades of development, stochastic dynamics has now become a relatively mature science field. The stochastic dynamics consists of two elements: the dynamical system which serves as a study object and whose input and output change over time; uncertainties that present in system properties, input and output which can be modeled as stochastic variables or stochastic process. Stochastic dynamics covers the following topics: modeling by characteristics of stochastic disturbance, methods, and procedures to obtain the stochastic response, stochastic stability, stochastic bifurcation, damage and reliability of the stochastic system. In brief, stochastic dynamics studies on qualitative and quantitative conditions of dynamical system under stochastic disturbances. This chapter introduces basic theories of stochastic dynamics.

2.1 Introduction

In many engineering branches, the modeling and analysis of dynamical system have always been a critical task. During the modeling process, considering uncertain changes of system parameters, disturbance changes, deviations from the modeling plan and other reasons, uncertainties are inevitable for a system. To calculate uncertainties more accurately, observation and measurement are usually made to get more data as much as possible. If there is an enough large amount of data for a specified uncertainty, the uncertainty can be described by probability and other statistics. In particular, if the uncertain physical quantity does not change with time, the uncertainty can be represented by a random variable. Moreover, if this uncertain physical quantity changes over time, the uncertainty can be modeled as a stochastic process. The following will give the basic theories of stochastic process, stochastic differential equation and stochastic system, which refer to the books [1–5].

P. Ju, *Stochastic Dynamics of Power Systems*, Power Systems,
https://doi.org/10.1007/978-981-13-1816-0_2

2.2 Stochastic Process

Considering a physical phenomenon that changes randomly over time, $X(t)$ is used to denote the physical quantity that needs to be studied in the stochastic phenomenon, and $X(t_1)$, $X(t_2)$, … are stochastic variables. This stochastic phenomenon can be studied by introducing a stochastic process, whose definition is: the stochastic process $X(t)$ is a family of stochastic variables that use t as a parameter from the index collection T, expressed by $\{X(t), t \in T\}$. Although an index collection can be of various types, only when it is a time collection, we can say $X(t)$ is a stochastic process in this book.

2.2.1 *Stochastic Process Description*

According to the definition, a stochastic process is a family of stochastic variables, which can be described stochastic variables jointly. The number of stochastic variables may be infinite, or even uncountable. However, a complete description of the stochastic process can be made only under a very limited actual condition. In most cases, the most important features, though incomplete, are sufficient for the analysis of applications such as engineering, biology, ecology, and finance.

1. **Probability Distribution**

As the situation of a stochastic variable, the probability function is only meaningful for the discrete stochastic process. However, a probability distribution function can be applicable to both discrete and continuous stochastic processes. It is more important that if the Dirac δ function is introduced, the probability density function can be used to describe both the discrete and continuous stochastic processes.

For example of a stochastic process $X(t)$, its first-order, second-order, and until n-order probability density functions are as follows:

$$p(x,t),\ p(x_1,t_1;x_2,t_2),\ \ldots\ldots, p(x_1,t_1;x_2,t_2;\ldots\ldots;x_n,t_n) \tag{2.2.1}$$

Because a lower-order probability density functions can be obtained from the integral of high order probability density functions, the higher order probability density functions contain more information than the lower order probability density functions.

2. **Moment Function**

A stochastic process can also be described by the following moment functions:

$$E[X(t)] = \int xp(x,t)\mathrm{d}x$$
$$E[X(t_1)X(t_2)] = \int\int x_1x_2p(x_1,t_1;x_2,t_2)\mathrm{d}x_1x_2$$
$$\cdots\cdots\cdots\cdots$$
$$E[X(t_1)X(t_2)\ldots\ldots X(t_n)] = \int\ldots\int x_1x_2\ldots x_np(x_1,t_1;x_2,t_2;\ldots;x_n,t_n)\mathrm{d}x_1x_2x_2\ldots x_n \tag{2.2.2}$$

The first-order and second-order moment functions are called mean value function and autocorrelation function, respectively, as follows:

$$\mu_X(t) = E[X(t)], \quad R_{XX}(t_1,t_2) = E[X(t_1)X(t_2)] \tag{2.2.3}$$

3. **Spectrum Description**

The autocorrelation function is the second-order statistical property of the stochastic process, because it involves two different times and is derived from the second-order probability density function. Another second-order statistical property equivalent to autocorrelation function is power spectrum density. It is one of the most important characteristic quantities for the stochastic process in practice.

This section only considers the zero-mean stationary stochastic process. Furthermore, the autocorrelation function is equivalent to the autocovariance function, which only depends on the time delay.

Consider the zero-mean stationary stochastic process $X(t)$. The power spectrum density function of $X(t)$ is defined as the Fourier transform of its autocorrelation function, which is as follows:

$$\Phi_{XX}(\omega) = \frac{1}{2\pi}\int_{-\infty}^{\infty} R_{XX}(\tau)e^{-i\omega\tau}\mathrm{d}\tau \tag{2.2.4}$$

Because $R_{XX}(\tau)$ is nonnegative definite, its Fourier transform $\Phi_{XX}(\omega)$ is also nonnegative definite. The inverse of (2.2.4) is

$$R_{XX}(\tau) = \int_{-\infty}^{\infty} \Phi_{XX}(\omega)e^{i\omega\tau}\mathrm{d}\omega \tag{2.2.5}$$

The Fourier transform pair (2.2.4) and (2.2.5) are known as the famous Wiener-Khinchin theorem.

To reveal the physical meaning of power spectrum density, $\tau = 0$ is assumed in (2.2.5), then

$$R_{XX}(0) = E[X^2(t)] = \int_{-\infty}^{\infty} \Phi_{XX}(\omega)\mathrm{d}\omega \tag{2.2.6}$$

Equation (2.2.6) shows, $\Phi_{XX}(\omega)$ describes how the mean-square value distributes in the whole frequency domain. In many cases, the mean-square value is the measurement of energy. For example, if $X(t)$ is the displacement of the mechanical system, then $X^2(t)$ is similar to the potential energy. In this situation, the power spectrum density $\Phi_{XX}(\omega)$ indicates how the energy distributes in the frequency domain. Therefore, the power spectrum density is also known as mean-square spectrum density.

2.2.2 *Gaussian Stochastic Process*

The stochastic process $X(t)$ is a family of stochastic variables that use time t as a parameter. If stochastic variables at all different time follow the Gaussian distribution, this stochastic process can be regarded as the Gaussian distribution. The stochastic variable following the Gaussian distribution can be defined by probability density function, so the Gaussian stochastic process can also be defined by probability density function.

The first-order probability density function of the Gaussian stochastic process is

$$p(x,t) = \frac{1}{\sqrt{2\pi}\,\sigma_X(t)} \exp\left\{-\frac{[x-\mu_X(t)]^2}{2\sigma_X^2(t)}\right\} \tag{2.2.7}$$

The statistic property of the probability density function with more than two orders can be obtained from the joint probability density function of Gaussian distribution variables. For example, the second-order probability density function is

$$\begin{aligned} p(x_1,t_1;x_2,t_2) = {} & \frac{1}{2\pi\sigma_1\sigma_2\sqrt{1-\rho^2}} \\ & \times \exp\left[-\frac{\sigma_2^2(x_1-\mu_1)^2 - 2\sigma_1\sigma_2\rho(x_1-\mu_1)(x_2-\mu_2) + \sigma_1^2(x_2-\mu_2)^2}{2\sigma_1^2\sigma_2^2(1-\rho^2)}\right] \end{aligned} \tag{2.2.8}$$

where $\mu_1 = \mu_X(t_1)$, $\mu_2 = \mu_X(t_2)$, $\sigma_1 = \sigma_X(t_1)$, $\sigma_2 = \sigma_X(t_2)$, and $\rho = \rho_{XX}(t_1,t_2)$.

If the Gaussian stochastic process is weakly stationary, the mean value function $\mu_X(t)$ is constant and the covariance function $\kappa_{XX}(t_1,t_2)$ only depends on time delay $\tau = t_2 - t_1$.

2.3 Stochastic Differential Equation

2.3.1 *Markov Process*

In the stochastic dynamics, the Markov process is a very important kind of processes. The reasons are: (i) it can be used as a model of many practical stochastic processes; (ii) the existing mathematical theory of Markov process can be used to solve many difficult stochastic problems; (iii) Markov process is easy to produce and simulate.

A stochastic process only has a temporary memory, so the current state can only be affected by the recent data. This type of process is collectively called as the Markov process. To be specific, a stochastic process $X(t)$ is known as the Markov process, if its conditional probability satisfies

$$\mathrm{Prob}[X(t_n) \le x_n | X(t_{n-1}) \le x_{n-1}, \ldots, X(t_1) \le x_1] = \mathrm{Prob}[X(t_n) \le x_n | X(t_{n-1}) \le x_{n-1}] \tag{2.3.1}$$

where $t_1 < t_2 < \ldots < t_n$. Obviously, the reason why the stochastic process $X(t)$ satisfies sufficient conditions for the Markov process is that the increments of the stochastic process in two non-overlapping time intervals are independent. That is to say, if $t_1 < t_2 \le t_3 < t_4$, then $X(t_2) - X(t_1)$ is independent of $X(t_4) - X(t_3)$. If $X(t)$ is a Gaussian process, the sufficient condition is that the two increments are not relevant, as follows

$$E\{[X(t_2) - X(t_1)][X(t_4) - X(t_3)]\} = 0, \quad t_1 < t_2 \le t_3 < t_4 \tag{2.3.2}$$

Obviously, the Markov process is just an ideal mathematic model of the real stochastic process. Even so, many stochastic processes can be represented by the Markov process. In physics, the Brownian motion is a Markov process. In many fields such as engineering, communication, ecology, and biology, many noise and signal processes are often modeled as the Markov process.

2.3.2 *Fokker-Planck-Kolmogorov Process*

Taking three moments $t_1 < t < t_3$ into account, one obtains

$$p(\mathbf{x}_2, t_2; \mathbf{y}, t | \mathbf{x}_1, t_1) = p(\mathbf{x}_2, t_2 | \mathbf{y}, t) p(\mathbf{y}, t | \mathbf{x}_1, t_1) \tag{2.3.3}$$

Integrating $\mathbf{y}$ in (2.3.3), one obtains

$$p(\mathbf{x}_2, t_2 | \mathbf{x}_1, t_1) = \int p(\mathbf{x}_2, t_2 | \mathbf{y}, t) p(\mathbf{y}, t | \mathbf{x}_1, t_1) \mathrm{d}\mathbf{y} \tag{2.3.4}$$

Equation (2.3.4) is the famous Chapman-Kolmogorov-Smoluchowski equation. It is an integral equation that dominates the transition probability density function.

To facilitate the analysis, an integral Eq. (2.3.4) can be transformed into an equivalent differential equation, i.e., the well-known Fokker-Planck-Kolmogorov (FPK) equation:

$$\frac{\partial}{\partial t}p+\sum_{j=1}^{n}\frac{\partial}{\partial x_j}(a_j p)-\frac{1}{2}\sum_{j,k=1}^{n}\frac{\partial^2}{\partial x_j \partial x_k}(b_{jk}p)+\frac{1}{3!}\sum_{j,k,l=1}^{n}\frac{\partial^3}{\partial x_j \partial x_k \partial x_l}(c_{jkl}p)-\cdots\cdots=0 \tag{2.3.5}$$

where $p = p(\mathbf{x}, t|\mathbf{x}_0, t_0)$ is the transition probability density function.

In many practical problems, the derivative functions with more than two orders are zero, so (2.3.5) becomes as

$$\frac{\partial}{\partial t}p+\sum_{j=1}^{n}\frac{\partial}{\partial x_j}(a_j p)-\frac{1}{2}\sum_{j,k=1}^{n}\frac{\partial^2}{\partial x_j \partial x_k}(b_{jk}p)=0 \tag{2.3.6}$$

The general FPK equation is (2.3.6). Meanwhile, the Markov process $\mathbf{X}(t)$ is known as a Markov diffusion process, or a diffusion process for short.

The FPK equation (2.3.6) could be rewritten as

$$\frac{\partial}{\partial t}p+\sum_{j=1}^{n}\frac{\partial}{\partial x_j}G_j=0 \tag{2.3.7}$$

where

$$G_j = a_j p-\frac{1}{2}\sum_{k=1}^{n}\frac{\partial}{\partial x_k}(b_{jk}p) \tag{2.3.8}$$

Equation (2.3.7) can be analogous to a continuous equation which indicates the fluid mass conservation in fluid mechanics, so (2.3.7) can be explained as a probability conservation equation, while G_j is the jth component of the probability current vector $G(\mathbf{x}, t|\mathbf{x}_0, t_0)$.

Equation (2.3.6) is a partial differential equation of the first-order time t and second-order state variable $\mathbf{x}$. When it comes to practical problems, the derivative moment $a_j(\mathbf{x},t)$ and $b_{jk}(\mathbf{x},t)$ can be determined by a system motion equation. In addition, to solve the FPK equation, it still needs derivation of the appropriate initial condition and boundary conditions according to the practical problems. Among many questions, the initial state is fixed, and the initial condition is

$$p(\mathbf{x}, t_0|\mathbf{x}_0, t_0) = \delta(\mathbf{x}-\mathbf{x}_0) = \prod_{j=1}^{n}\delta(x_j - x_{j0}) \tag{2.3.9}$$

The boundary condition depends on sample characteristics of the system. For the non-infinite boundaries, there are several classical types: reflecting boundary, absorbing boundary and periodic boundary. However, for many engineering questions, infinite boundaries are very important. For example of an infinite boundary, its probability current is zero, which is as follows

$$\lim_{x_j \to \pm\infty} \mathbf{G}(\mathbf{x}, t|\mathbf{x}_0, t_0) = 0 \tag{2.3.10}$$

Moreover, because the total probability is limited, one obtains

$$\lim_{x_j \to \pm\infty} p(\mathbf{x}, t|\mathbf{x}_0, t_0) = 0 \tag{2.3.11}$$

Meanwhile, it goes toward zero like $|x_j|^{-\alpha}$ ($\alpha > 1$), depending on the specific system.

When the Markov diffusion process is approaching a stationary state, its stationary probability density function becomes the limit of the transition probability density function. Meanwhile, the time derivative terms in (2.3.6) are zero, thus one obtains a so-called simplified FPK equation:

$$\sum_{j=1}^{n} \frac{\partial}{\partial x_j}(a_j p) - \frac{1}{2}\sum_{j,k=1}^{n} \frac{\partial^2}{\partial x_j \partial x_k}(b_{jk} p) = 0 \tag{2.3.12}$$

where $p = p(\mathbf{x})$ is the stationary probability density function, and a_j and b_{jk} are time-independent. Equation (2.3.12) can be rewritten as

$$\sum_{j=1}^{n} \frac{\partial}{\partial x_j} G_j = 0,\ G_j = a_j p - \frac{1}{2}\sum_{k=1}^{n} \frac{\partial}{\partial x_k}(b_{jk} p) \tag{2.3.13}$$

2.3.3 Kolmogorov's Backward Equation

In an FPK equation, an unknown term $p(\mathbf{x}, t|\mathbf{x}_0, t_0)$ is regarded as a function of t and $\mathbf{x}$, and t_0 and $\mathbf{x}_0$ are regarded as parameters. The FPK equation is also known as the Kolmogorov's Forward Equation, because the term $\frac{\partial p}{\partial t}$ is a derivative of the later time t, and it is known as a forward variable together with a state variable $\mathbf{x}$ corresponding to the later time t. Conversely, $p(\mathbf{x}, t|\mathbf{x}_0, t_0)$ can also be regarded as a function of t_0 and $\mathbf{x}_0$, while t and $\mathbf{x}$ are regarded as parameters. The other equation that corresponds to (2.3.5) is

$$\frac{\partial p}{\partial t_0} + \sum_{j=1}^{n} a_j \frac{\partial p}{\partial x_{j0}} + \frac{1}{2}\sum_{j,k=1}^{n} b_{jk} \frac{\partial^2 p}{\partial x_{j0} \partial x_{k0}} + \frac{1}{3!}\sum_{j,k,l=1}^{n} c_{jkl} \frac{\partial^3 p}{\partial x_{j0} \partial x_{k0} \partial x_{l0}} + \cdots\cdots = 0 \tag{2.3.14}$$

where a_j, b_{jk}, c_{jkl}, … are still derivative moments, but they are functions of $\mathbf{x}_0$ and t_0.

For the Markov diffusion process, (2.3.14) is transformed into

$$\frac{\partial p}{\partial t_0}+\sum_{j=1}^{n}a_j(\mathbf{x}_0,t_0)\frac{\partial p}{\partial x_{j0}}+\frac{1}{2}\sum_{j,k=1}^{n}b_{jk}(\mathbf{x}_0,t_0)\frac{\partial^2 p}{\partial x_{j0}\partial x_{k0}}=0 \tag{2.3.15}$$

Equations (2.3.14) and (2.3.15) are known as Kolmogorov's backward equations, and $\mathbf{x}_0$ is a backward variable. The FPK equation (forward equation) is often used to obtain the probability density function, while the backward equation can be used to study the first passage issue.

2.3.4 *Wiener Process*

The simplest Markov diffusion process is the Wiener process, also known as the Brownian motion, which is represented by $B(t)$. A stochastic process $B(t)$ can be called as the Wiener process, as long as the following requirements are met: (i) $B(t)$ is a Gaussian process, (ii) $B(0) = 0$, (iii) $E[B(t)] = 0$, and (iv)

$$E[B(t_1)B(t_2)]=\sigma^2\min(t_1,t_2) \tag{2.3.16}$$

where σ^2 is called the intensity of a Wiener process. Equation (2.3.16) implies that the Wiener process is not a stationary process. Given that $t_1 < t_2 \leq t_3 < t_4$, one obtains the following equation from (2.3.16)

$$\begin{aligned}&E\{[B(t_2)-B(t_1)][B(t_4)-B(t_3)]\}\\&=E[B(t_2)B(t_4)-B(t_1)B(t_4)-B(t_2)B(t_3)+B(t_1)B(t_3)]\\&=\sigma^2(t_2-t_1-t_2+t_1)=0\end{aligned} \tag{2.3.17}$$

Based on the sufficient condition (2.3.2), $B(t)$ is a Markov process.

The relationship between the Wiener process and the Gaussian white noise is

$$\frac{\mathrm{d}B(t)}{\mathrm{d}t}=W(t) \tag{2.3.18}$$

and the relationship between the Wiener process's intensity and spectral density is

$$\sigma^2=2\pi K \tag{2.3.19}$$

where σ^2 is the Wiener process's intensity.

2.3.5 Itô Stochastic Differential Equation

As the simplest Markov diffusion process, the Wiener process $B(t)$ can be applied to establish other Markov diffusion processes through a stochastic differential equation. According to Itô, a scalar Markov diffusion process can be produced by

$$\mathrm{d}X(t) = m(X,t)\mathrm{d}t + \sigma(X,t)\mathrm{d}B(t) \tag{2.3.20}$$

where $B(t)$ is a unit Wiener process, which is

$$E[B(t_1)B(t_2)] = \min(t_1,t_2), \quad E[\mathrm{d}B(t_1)\mathrm{d}B(t_2)] = \begin{cases} 0, \ t_1 \neq t_2 \\ \mathrm{d}t, \ t_1 = t_2 = t \end{cases} \tag{2.3.21}$$

In (2.3.20), terms m and σ depend on $X(t)$, and can also explicitly include t. Equation (2.3.20) has the following solution:

$$X(t) = X(0) + \int_0^t m(X,\tau)\mathrm{d}\tau + \int_0^t \sigma(X,\tau)\mathrm{d}B(\tau) \tag{2.3.22}$$

The last term in (2.3.20) is a Stieltjes integral, which is

$$\int_0^t \sigma[X(\tau),\tau]\mathrm{d}B(\tau) = \underset{\substack{n\to\infty \\ \Delta_n\to 0}}{\mathrm{l.i.m.}} \sum_{j=1}^{n} \sigma[X(\tau_j'),\tau_j'][B(\tau_{j+1}) - B(\tau_j)] \tag{2.3.23}$$

where $0 = \tau_1 < \tau_2 < \ldots < \tau_n < \tau_{n+1} = t$, $\Delta_n = \max(\tau_{j+1} - \tau_j)$, and $\tau_j \le \tau_j' \le \tau_{j+1}$. Since $B(t)$ is a very unusual stochastic process, it is non-differentiable and has an unbounded change at any limited time interval. This Stieltjes integral must be explained appropriately, with its key point being the selection of τ_j'. There were two choices proposed before: one from Itô and another from Stratonovich.

Itô integral chooses $\tau_j' = \tau_j$ while Stieltjes integral (2.3.23) becomes as

$$\int_0^t \sigma[X(\tau),\tau]\mathrm{d}B(\tau) = \underset{\substack{n\to\infty \\ \Delta_n\to 0}}{\mathrm{l.i.m.}} \sum_{j=1}^{n} \sigma[X(\tau_j),\tau_j][B(\tau_{j+1}) - B(\tau_j)] \tag{2.3.24}$$

With Itô integral, the differential equation (2.3.20) is known as the Itô stochastic differential equation. The stochastic process $X(t)$ is a diffusion process, and functions m and σ are called a shift coefficient and a diffusion coefficient, respectively. In Itô integral (2.3.24), a difference $B(\tau_{j+1}) - B(\tau_j)$ takes its value from a forward time interval of τ_j, while $\sigma[X(\tau), \tau]$ gets its value from τ_j, which ensures $\mathrm{d}B(t)$ in (2.3.20) is independent of $X(t)$.

To obtain the first-order and second-order derivative moments, consider a very small Δt, written as

$$\begin{aligned} X(t+\Delta t) - X(t) &= \int_{t}^{t+\Delta t} m[X(\tau),\tau]\mathrm{d}\tau + \int_{t}^{t+\Delta t} \sigma[X(\tau),\tau]\mathrm{d}B(\tau) \\ &\approx m[X(t),t]\Delta t + \sigma[X(t),t][B(t+\Delta t) - B(t)] \end{aligned} \tag{2.3.25}$$

$$a(x,t) = \lim_{\Delta t \to 0} \frac{1}{\Delta t} E[X(t+\Delta t) - X(t)|X(t) = x] = m(x,t) \tag{2.3.26}$$

$$b(x,t) = \lim_{\Delta t \to 0} \frac{1}{\Delta t} E\{[X(t+\Delta t) - X(t)]^2|X(t) = x\} = \sigma^2(x,t) \tag{2.3.27}$$

Therefore, the first-order and second-order derivative moments in the FPK equation can be directly obtained from the shift coefficient and the diffusion coefficient. However, it should be noted that the first-order and second-order derivative moments are functions of the state variable x, while the shift coefficient and diffusion coefficient are functions of the stochastic process $X(t)$.

The above analysis can also be extended to an n-dimensional case. A n-dimensional Markov diffusion vector process is produced from the Itô stochastic differential equation set below:

$$\mathrm{d}X_j(t) = m_j(\mathbf{X},t)\mathrm{d}t + \sum_{l=1}^{m} \sigma_{jl}(\mathbf{X},t)\mathrm{d}B_l(t), \quad j = 1,2,\ldots,n \tag{2.3.28}$$

where $B_l(t)$, l = 1, 2, …, m, which constitute an independent unit Wiener process. The first-order and second-order derivative moments of a proper FPK equation are given by the following equation:

$$a_j(\mathbf{x},t) = m_j(\mathbf{x},t),\ b_{jk}(\mathbf{x},t) = \sum_{l=1}^{m} \sigma_{jl}(\mathbf{x},t)\sigma_{kl}(\mathbf{x},t) \tag{2.3.29}$$

It can be written in a matrix form:

$$\mathbf{a}(\mathbf{x},t) = \mathbf{m}(\mathbf{x},t),\ \mathbf{b}(\mathbf{x},t) = \boldsymbol{\sigma}(\mathbf{x},t)\boldsymbol{\sigma}^{\mathrm{T}}(\mathbf{x},t) \tag{2.3.30}$$

Considering one function $F(\mathbf{X},t)$ of a Markov vector process $\mathbf{X}$(t), this function is differentiable to t, and twice differentiable to $\mathbf{X}(t)$. The differential of $F(\mathbf{X},t)$ is

$$\mathrm{d}F(\mathbf{X},t) = \frac{\partial F}{\partial t}\mathrm{d}t + \sum_{j=1}^{n} \frac{\partial F}{\partial X_j}\mathrm{d}X_j + \frac{1}{2}\sum_{j,k=1}^{n} \frac{\partial^2 F}{\partial X_j \partial X_k}\mathrm{d}X_j\mathrm{d}X_k + \ldots \tag{2.3.31}$$

Perform dX_j of (2.3.28) to use properties of the Wiener process, as well as dt and $dB_l(t)$ order amounts, which gives

$$dF(\mathbf{X},t) = \left(\frac{\partial F}{\partial t} + \sum_{j=1}^{n} m_j \frac{\partial F}{\partial X_j} + \frac{1}{2}\sum_{j,k=1}^{n}\sum_{l=1}^{m} \sigma_{jl}\sigma_{kl}\frac{\partial^2 F}{\partial X_j \partial X_k}\right)dt + \sum_{j=1}^{n}\sum_{l=1}^{m}\sigma_{jl}\frac{\partial F}{\partial X_j}dB_l \tag{2.3.32}$$

Equation (2.3.32) is known as the Itô differential rule or Itô lemma. For a one-dimensional system subject to a single Wiener process disturbance, (2.3.32) is simplified into

$$dF(X,t) = \left(\frac{\partial F}{\partial t} + m\frac{dF}{dX} + \frac{1}{2}\sigma^2\frac{d^2F}{dX^2}\right)dt + \sigma\frac{dF}{dX}dB(t) \tag{2.3.33}$$

Itô lemma shows that the Itô equation of a Markov diffusion process function can be derived directly and easily. This property is one advantage from using the Itô stochastic differential equation to describe the Markov diffusion process.

2.4 Stochastic System

It is difficult to get an exact solution for a system subject to stochastic disturbances. However, when stochastic disturbances are idealized as Gaussian white noises and the system response as the Markov diffusion process, it is possible to obtain a precise solution. The probability density function of the Markov diffusion process is dominated by an FPK equation, which can be derived from system motion equations. Nevertheless, the complete solution of an FPK equation can only be obtained for the very special first-order system, which represents how the probabilistic structure evolves over time.

2.4.1 Dissipated Hamiltonian System Subject to Stochastic Disturbance

The nonlinear stochastic dynamical system can be referred to as the dissipated Hamiltonian system subject to stochastic disturbances, which is particularly available for dealing with multi-degree-of-freedom strong nonlinear stochastic dynamical systems. The systematic method of getting accurate and stable solutions for a dissipated Hamiltonian subject to stochastic disturbances has been developed.

Without the energy dissipation and disturbances, the motion equation of an n degree-of-freedom Hamiltonian system has a form of

$$\dot{q}_j = \frac{\partial H}{\partial p_j}, \dot{p}_j = -\frac{\partial H}{\partial q_j}, i = 1, 2, \ldots, n \tag{2.4.1}$$

where q_j and p_j are the generalized coordinate and the generalized momentum, respectively. Meanwhile, $H = H(\mathbf{q}, \mathbf{p})$ is a Hamiltonian function that has first-order partial derivatives. The Hamiltonian system can be classified into three types: completely integrable, partially integrable, and completely non-integrable. It is difficult to apply the classification criteria. Thus, discussions in this book only cover three simple and practical types.

Considering an n-degree-of-freedom Hamiltonian system, if there are n independent motion integrals $H_1, H_2, \ldots, H_n$, Hamiltonian functions can be represented as follows:

$$H(\mathbf{q}, \mathbf{p}) = \sum_{j=1}^{n} H_j(q_j, p_j) \tag{2.4.2}$$

Then, this Hamiltonian system belongs to the completely integrable type. If there are only $r + 1 (r + 1 < n)$ independent motion integrals, and the Hamiltonian function can be represented as

$$H(\mathbf{q}, \mathbf{p}) = \sum_{j=1}^{r} H_j(q_j, p_j) + \tilde{H}(q_{r+1}, q_{r+2,\ldots,} q_n; p_{r+1,} p_{r+2,\ldots,} p_n) \tag{2.4.3}$$

This Hamiltonian system belongs to the partially integrable one. For extreme cases, if there is only one motion integral of a Hamiltonian function, i.e., $r = 0$, the Hamiltonian system is completely non-integrable.

For the completely or partially integrable Hamiltonian systems, if there is a rational relationship between the two motion frequencies, (e.g., if the two frequencies are equal or one is twice of the other), an internal resonance may be induced between the two motions. Therefore, each of these two types of Hamiltonian systems can also be sorted into two subclasses: non-resonant and resonant. Consequently, the Hamiltonian system consists of five types: completely non-integrable, completely integrable and non-resonant, completely integrable and resonant, partial integrable and non-resonant, as well as partial integrable and resonant. It should be mentioned that there is no internal resonance in completely non-integrable situations, because it only involves one motion in all dimensions.

2.4.2 *Stochastic Averaging Method*

As previously mentioned, disturbances are mostly required to be Gaussian white noises, so accurate or approximate solutions for the system response of the Markov diffusion process can be obtained. It is known that white noise is an ideal mathematic process whose correlative function is the δ function. Actual physical processes cannot satisfy this requirement. In the following, a method will be developed to use white noises to approximate non-white noises. Feasible conditions of this approximation will be studied accordingly.

Given that the system motion equation is

$$\frac{\mathrm{d}}{\mathrm{d}t}X_j(t) = f_j(\mathbf{X}, t) + \sum_{l=1}^{m} g_{jl}(\mathbf{X}, t)\xi_l(t), \quad j = 1, 2, \ldots, n \tag{2.4.4}$$

where $\xi_l(t)$ is a zero-mean joint stationary process, which has the following correlative functions:

$$E[\xi_l(u)\xi_s(v)] = R_{ls}(\tau), \ \tau = v - u \tag{2.4.5}$$

The related time between $\xi_l(t)$ and $\xi_s(t)$ is defined as

$$\tau_{ls} = \frac{1}{\sqrt{R_{ll}(0)R_{ss}(0)}} \int_{-\infty}^{0^-} |R_{ls}(\tau)| \mathrm{d}\tau \tag{2.4.6}$$

It is a memory measurement of the current $\xi_s(t)$ to the past $\xi_l(t - \tau)$. For the white noise, it is zero. If all the disturbances $\xi_l(t)$ could be approximated as the Gaussian white noise, the probability density function of the system response $\mathbf{X}$ could satisfy the FPK equation. Thus, the main target of approximation is to determine the first-order and second-order derivative moments a_j and b_{jk} by the original system Eq. (2.4.4). A time interval of two close measurements is regarded as Δt. In order to get a_j, the following increment is calculated

$$X_j(t + \Delta t) - X_j(t) = \int_t^{t+\Delta t} f_j(\mathbf{X}_u, u)\mathrm{d}u + \sum_{l=1}^{m} \int_t^{t+\Delta t} g_{jl}(\mathbf{X}_u, u)\xi_l(u)\mathrm{d}u \tag{2.4.7}$$

where $\mathbf{X}_u$ is the abbreviation of $\mathbf{X}(u)$. f_j and g_{jl} in (2.4.7) are expanded at time t, and then one obtains

$$\begin{aligned} f_j(\mathbf{X}_u, u) =& f_j(\mathbf{X}_t, t) + (u - t)\frac{\partial}{\partial t}f_j(\mathbf{X}_t, t) \\ &+ \sum_{r=1}^{n} [X_r(u) - X_r(t)]\frac{\partial}{\partial X_r}f_j(\mathbf{X}_t, t) + \ldots \end{aligned} \tag{2.4.8}$$

$$\begin{aligned} g_{jl}(\mathbf{X}_u, u) =& g_{jl}(\mathbf{X}_t, t) + (u - t)\frac{\partial}{\partial t}g_{jl}(\mathbf{X}_t, t) \\ &+ \sum_{r=1}^{n} [X_r(u) - X_r(t)]\frac{\partial}{\partial X_r}g_{jl}(\mathbf{X}_t, t) + \cdots \end{aligned} \tag{2.4.9}$$

In (2.4.8) and (2.4.9), $X_r(u) - X_r(t)$ is replaced by

$$X_r(u) - X_r(t) = \int_t^u f_r(\mathbf{X}_v, v)\mathrm{d}v + \sum_{s=1}^{m} \int_t^u g_{rs}(\mathbf{X}_v, v)\xi_s(v)\mathrm{d}v \tag{2.4.10}$$

Combining Equation (2.4.7) into (2.4.10) and keeping main items, one obtains

$$\begin{aligned} X_j(t+\Delta t) - X_j(t) = & \int_t^{t+\Delta t} f_j(\mathbf{X}_t, t)\mathrm{d}u + \sum_{l=1}^{m} \int_t^{t+\Delta t} g_{jl}(\mathbf{X}_t, t)\xi_l(u)\mathrm{d}u \\ & + \sum_{l=1}^{m} \int_t^{t+\Delta t} (u-t)\left[\frac{\partial}{\partial t} g_{jl}(\mathbf{X}_t, t)\right]\xi_l(u)\mathrm{d}u \\ & + \sum_{l,s=1}^{m} \int_t^{t+\Delta t} \xi_l(u)\mathrm{d}u \sum_{r=1}^{n} \frac{\partial}{\partial X_r} g_{jl}(\mathbf{X}_t, t) \int_t^u g_{rs}(\mathbf{X}_v, v)\xi_s(v)\mathrm{d}v \end{aligned} \tag{2.4.11}$$

Based on (2.4.11), one could obtain

$$\begin{aligned} a_j(\mathbf{x}_t, t) = & f_j(\mathbf{x}_t, t) + \frac{1}{\Delta t}\sum_{l,s=1}^{m}\sum_{r=1}^{n} \int_t^{t+\Delta t} \mathrm{d}u \int_t^u \left[\frac{\partial}{\partial x_r} g_{jl}(\mathbf{x}_t, t)\right] g_{rs}(\mathbf{x}_v, v) \\ & E[\xi_l(u)\xi_s(v)]\mathrm{d}v + 0(\Delta t) \end{aligned} \tag{2.4.12}$$

where $0(\Delta t)$ denotes the high-order remainder term of Δt. In the derivation of (2.4.12), it is assumed that the variation of f_j and g_{jl} functions at Δt time interval is not significantly affected by the stochastic property of $\mathbf{X}(t)$. Therefore, they are excluded from the ensemble average. Substituting (2.4.5) into (2.4.12), and changing the integral variable from v to $\tau = v\text{-}u$, one obtains

$$\begin{aligned} a_j(\mathbf{x}_t, t) = & f_j(\mathbf{x}_t, t) + \frac{1}{\Delta t}\sum_{l,s=1}^{m}\sum_{s=1}^{n} \int_t^{t+\Delta t} \mathrm{d}u \int_{t-u}^{0} \left[\frac{\partial}{\partial x_r} g_{jl}(\mathbf{x}_t, t)\right] g_{rs}(\mathbf{x}_{u+\tau}, u+\tau) \\ & R_{ls}(\tau)\mathrm{d}\tau + 0(\Delta t) \end{aligned} \tag{2.4.13}$$

If Δt is much longer than the related time τ_{ls}, i.e., performing $\tau > \Delta t$, $R_{ls}(\tau) \approx 0$, and then changing the integration order (i.e., u is integrated first), one can rewrite (2.4.13) as

$$a_j(\mathbf{x}_t, t) = f_j(\mathbf{x}_t, t) + \sum_{l,s=1}^{m} \sum_{r=1}^{n} \int_{-\Delta t}^{0} \left[\frac{\partial}{\partial x_r} g_{jl}(x_t, t) \right] g_{rs}(x_{t+\tau}, t+\tau) R_{ls}(\tau) \mathrm{d}\tau \quad (2.4.14)$$

For a second-order derivative moment, (2.4.7) is substituted into (2.4.14) and the similar steps are taken. Then, one obtains

$$b_{jk}(\mathbf{x}_t, t) = \sum_{l,s=1}^{m} \int_{-\Delta t}^{\Delta t} g_{jl}(\mathbf{x}_t, t) g_{ks}(\mathbf{x}_{t+\tau}, t+\tau) R_{ls}(\tau) \mathrm{d}\tau \quad (2.4.15)$$

In the above derivation, it is assumed that the functions f_j and g_{jl} are slowly changing in the interval Δt. To satisfy this condition, Δt could be required no longer than the relaxation time of the system, which is a measure of the system motion changing rate without disturbances. The relaxation time τ_{rel} has been defined for oscillating and non-oscillating systems. For an oscillating system, τ_{rel} is the time that vibration amplitude decreases to e^{-1} times or increases it to e times. In contrast, for a non-oscillating system, it is the time that the amplitude enlarges to e times. If Δt is greater than τ_{rel}, the functions f_j and g_{jl} have significant changes, which may lose many details.

From the derivation (2.4.14) and (2.4.15), it is known that the time interval Δt must be significantly longer than the related time but shorter than the relaxation time. Thus, the condition of (2.4.14) and (2.4.15) is that the relaxation time of the system is much longer than the relevant time of all the disturbances. In this case, the system's response could be approximated to the Markov diffusion process.

Since Δt is much longer than the related time of disturbances, the correlation function can only be non-zero in a small neighborhood of $\tau = 0$. As a result, the lower limit of the integrals in (2.4.14) and (2.4.15) can be extended to $-\infty$, while the upper limit of the integrals in (2.4.15) can be extended to ∞, which are

$$a_j(\mathbf{x}_t, t) = f_j(\mathbf{x}_t, t) + \sum_{l,s=1}^{m} \sum_{s=1}^{n} \int_{-\infty}^{0} \left[\frac{\partial}{\partial x_r} g_{jl}(\mathbf{x}_t, t) \right] g_{rs}(\mathbf{x}_{t+\tau}, t+\tau) R_{ls}(\tau) \mathrm{d}\tau \quad (2.4.16)$$

$$b_{jk}(\mathbf{x}_t, t) = \sum_{l,s=1}^{m} \int_{-\infty}^{\infty} g_{jl}(\mathbf{x}_t, t) g_{ks}(\mathbf{x}_{t+\tau}, t+\tau) R_{ls}(\tau) \mathrm{d}\tau \quad (2.4.17)$$

The above two equations can be used for calculating the approximate first-order and second-order derivative moments in the FPK equation. They are can be used for various stochastic averages, as well as approximating a white noise to a non-white noise and approximating a system response to the Markov diffusion process.

In the stationary stochastic process of a δ-relevant disturbance (i.e., the special case of white noise), one has

$$E[\xi_l(u)\xi_s(v)] = R_{ls}(v-u) = 2\pi K_{ls}\delta(v-u) \tag{2.4.18}$$

where K_{ls} is a constant spectral density. Equations (2.4.16) and (2.4.17) are simplified to (2.4.19) and (2.4.20), respectively, which are also shown as follows:

$$a_j(\mathbf{x}_t, t) = f_j(\mathbf{x}_t, t) + \sum_{l,s=1}^{m}\sum_{r=1}^{n} \pi K_{ls} g_{rs}(\mathbf{x}_t, t)\frac{\partial}{\partial x_r} g_{jl}(\mathbf{x}_t, t) \tag{2.4.19}$$

$$b_{jk}(\mathbf{x}_t, t) = \sum_{l,s=1}^{m} 2\pi K_{ls} g_{jl}(\mathbf{x}_t, t) g_{ks}(\mathbf{x}_t, t) \tag{2.4.20}$$

The second term in the right of (2.4.19) is a Wong-Zakai correction term.

Another step of the stochastic averaging method is the time averaging. The time averaging is used in two cases. One is that the functions f_j and g_{jl} in (2.4.19) and (2.4.20) are periodic functions with period T_p. Furthermore, the time averaging can be made in one period, as follows

$$\langle[\cdot]\rangle_t = \lim_{T\to\infty}\frac{1}{2T}\int_{-T}^{T}[\cdot]\mathrm{d}t = \frac{1}{T_p}\int_{0}^{T_p}[\cdot]\mathrm{d}t \tag{2.4.21}$$

It is convenient to use the Itô stochastic differential equation to analyze the Markov diffusion process. The average Itô equation is given by (2.3.28), and its drift and diffusion coefficients are obtained by time averages of (2.4.16) and (2.4.17) as follows:

$$m_j(\mathbf{X}) = \langle f_j(\mathbf{X}_t, t)\rangle_t + \sum_{l,s=1}^{m}\sum_{r=1}^{n_1}\int_{-\infty}^{0}\left\langle g_{rs}(\mathbf{X}_{t+\tau}, t+\tau)\frac{\partial}{\partial X_r} g_{jl}(\mathbf{X}_t, t)\right\rangle_t R_{ls}(\tau)\mathrm{d}\tau \tag{2.4.22}$$

$$\left(\boldsymbol{\sigma}\boldsymbol{\sigma}^T\right)_{jk} = \sum_{r=1}^{n_1}\sigma_{jr}(\mathbf{X})\sigma_{kr}(\mathbf{X}) = \sum_{l,s=1}^{m}\int_{-\infty}^{\infty}\langle g_{jl}(\mathbf{X}_t, t) g_{ks}(\mathbf{X}_{t+\tau}, t+\tau)\rangle_t R_{ls}(\tau)\mathrm{d}\tau \tag{2.4.23}$$

In (2.4.22) and (2.4.23), the average drift and diffusion systems are smoothed without the explicit time term. After the time averaging, the total response vector of the average stochastic system is still a vector diffusion process, whose dimension is the same as that of the original system. However, it is also possible that the time-averaged single response (or sub-vector of the response vector) itself is a diffusion process (or a vector diffusion process). As a result, the system dimension could be reduced. The method of obtaining (2.4.22) and (2.4.23) were proposed by Stratonovich, which were called 'stochastic averaging' in the literature.

For the situation of Gaussian white noise disturbances, the average drift and diffusion coefficients become

$$m_j(\mathbf{X}) = \langle f_j(\mathbf{X}_t, t)\rangle_t + \sum_{l,s=1}^{m}\sum_{r=1}^{n} \pi K_{ls}\left\langle g_{rs}(\mathbf{X}_t, t)\frac{\partial}{\partial x_r} g_{jl}(\mathbf{X}_t, t)\right\rangle_t \tag{2.4.24}$$

$$\left(\boldsymbol{\sigma}\boldsymbol{\sigma}^T\right)_{jk} = \sum_{l,s=1}^{m} 2\pi K_{ls}\langle g_{jl}(\mathbf{X}_t, t) g_{ks}(\mathbf{X}_t, t)\rangle_t \tag{2.4.25}$$

Another case of time averaging is that the system states can be sorted into fast-changing states and slow-changing states in terms of timescale. Without loss of generality, it is assumed that the first $n_1 (n_1 < n)$ state variables in the system (2.4.4) are slow-changing, while all other state variables are fast-changing. Therefore, the first n_1 equations become

$$\frac{\mathrm{d}}{\mathrm{d}t}X_j(t) = \varepsilon f_j(\mathbf{X}, t) + \varepsilon^{1/2}\sum_{l=1}^{m} g_{jl}(\mathbf{X}, t)\xi_l(t),\ j = 1, \ldots, n_1 \tag{2.4.26}$$

to replace the first n_1 equations in (2.4.4). The parameter $\varepsilon \ll 1$ is introduced to indicate that the first and second terms on the right are ε and $\varepsilon^{1/2}$ order small quantities, respectively, while $X_j(t)$, $j = 1, 2, \ldots, n_1$ is slow-changing. Furthermore, it is quickly known that this operation is equivalent to the assumption that the contribution of the right two items to the system is of the same order. After the stochastic averaging and time averaging, the following equation of $X_j(t)$ ($j = 1, 2, \ldots, n_1$) can be obtained.

$$\mathrm{d}X_j(t) = m_j(\tilde{\mathbf{X}})\mathrm{d}t + \sum_{l=1}^{m} \sigma_{jl}(\tilde{\mathbf{X}})\mathrm{d}B_l(t) \tag{2.4.27}$$

where $\tilde{\mathbf{X}} = [X_1, X_2, \cdots X_{n_1}]^T$,

$$m_j(\tilde{\mathbf{X}}) = \varepsilon\left\{\langle f_j(\mathbf{X}_t, t)\rangle_t + \sum_{l,s=1}^{m}\sum_{r=1}^{n_1}\int_{-\infty}^{0}\left\langle g_{rs}(\mathbf{X}_{t+\tau}, t+\tau)\frac{\partial}{\partial X_r} g_{jl}(\mathbf{X}_t, t)\right\rangle_t R_{ls}(\tau)\mathrm{d}\tau\right\} \tag{2.4.28}$$

$$\left(\boldsymbol{\sigma}\boldsymbol{\sigma}^{T}\right)_{jk}=\sum_{r=1}^{n_1}\sigma_{jr}(\tilde{\mathbf{X}})\sigma_{kr}(\tilde{\mathbf{X}})=\varepsilon\sum_{l,s=1}^{m}\int_{-\infty}^{\infty}\left\langle g_{jl}(\mathbf{X}_t,t)g_{ks}(\mathbf{X}_{t+\tau},t+\tau)\right\rangle_t R_{ls}(\tau)\mathrm{d}\tau \tag{2.4.29}$$

In time averaging, slow-changing variables are regarded to be constant. Fast-changing variables are changing stochastically with time so that it is difficult to use time averaging for these variables. Therefore, it is often based on the system characteristics, the space averaging of fast variables is used to substitute the time averaging. By averaging all the fast-changing variables and leaving only slow-changing variables, the vector Markov diffusion process could be constituted. Slow-changing variables reflect the general trend of the system motion. Consequently, the dimension of the system is reduced to that of slow variables, so problems are simplified.

The method of getting (2.4.27)–(2.4.29) is called the stochastic averaging of Itô equations with fast-changing and slow-changing variables. Obviously, there are two steps in the derivation of (2.4.22)–(2.4.25) or (2.4.27)–(2.4.29). The first one is to obtain the approximate white Gaussian noise as a disturbance process and the system response as a Markov diffusion process. The obtained equations (2.4.16) and (2.4.17) are non-smooth first-order and second-order derivative moments, and all variables are retained. The second one is to use time averaging. In the first case, all variables are retained. In the second case, fast-changing variables are eliminated to reduce the system dimension. The combinated method is called the smooth type of stochastic averaging method. The stochastic averaging method refers to both smooth type (2.4.22)–(2.4.25) or (2.4.27)–(2.4.29), and non-smooth type (2.4.16)–(2.4.17).

In this section, the stochastic averaging method is mainly applied to single-degree-of-freedom nonlinear stochastic systems. Also, it can be used for multi-degree-of-freedom nonlinear stochastic systems, but the steps are more complicated. The next section will introduce the quasi-Hamiltonian stochastic averaging method, which can be applied to multi-degree-of-freedom strongly-nonlinear stochastic systems.

2.4.3 Stochastic Averaging Method for Quasi Hamiltonian System

Consider a dissipative Hamiltonian system with stochastic disturbances governed by the following equations:

$$\dot{Q}_j=\frac{\partial H'}{\partial P_j},\quad \dot{P}_j=-\frac{\partial H'}{\partial Q_j}-\varepsilon\sum_{k=1}^{n}c_{jk}(\mathbf{Q},\mathbf{P})\frac{\partial H'}{\partial P_k}+\varepsilon^{1/2}\sum_{l=1}^{m}g_{jl}(\mathbf{Q},\mathbf{P})W_l(t) \tag{2.4.30}$$

where ε is a small parameter. $W_l(t)$ is Gaussian white noise with the following correlation function.

$$E[W_l(t)W_s(t+\tau)] = 2\pi K_{ls}\delta(\tau), \quad l, s = 1, 2, \ldots, m \tag{2.4.31}$$

The damping force and disturbance of this kind of system are small, so this kind of system is called the quasi Hamiltonian system. In this section, only a completely non-integrable Hamiltonian system is considered.

The Itô stochastic differential equations that equivalent to (2.4.30) are

$$\begin{aligned} \mathrm{d}Q_j &= \frac{\partial H'}{\partial P_j}\mathrm{d}t, \\ \mathrm{d}P_j &= \left[-\frac{\partial H'}{\partial Q_j} - \varepsilon\sum_{k=1}^{n} c_{jk}\frac{\partial H'}{\partial P_k} + \varepsilon\pi\sum_{k=1}^{n}\sum_{l,s=1}^{m} K_{js}g_{ks}\frac{\partial g_{jl}}{\partial P_k}\right]\mathrm{d}t + \varepsilon^{1/2}\sum_{l=1}^{m} g_{jl}\mathrm{d}B_l(t) \end{aligned} \tag{2.4.32}$$

where $B_l(t)$ is the Wiener process.

$$E[\mathrm{d}B_l(t)\mathrm{d}B_s(t+\tau)] = 2\pi K_{ls}\delta(\tau)\mathrm{d}t, \quad l, s = 1, 2, \ldots, m \tag{2.4.33}$$

The multi summations in the right of (2.4.32) are Wong-Zakai corrections that may affect resilience and damping forces. Under these influences, the original Hamiltonian H' may become H, the damping coefficient may change from c_{jk} to m_{jk}, and then the system Eq. (2.4.32) become

$$\mathrm{d}Q_j = \frac{\partial H}{\partial P_j}\mathrm{d}t, \quad \mathrm{d}P_j = -\left(\frac{\partial H}{\partial Q_j} - \varepsilon\sum_{k=1}^{n} m_{jk}\frac{\partial H}{\partial P_k}\right)\mathrm{d}t + \varepsilon^{1/2}\sum_{l=1}^{m} g_{jl}\mathrm{d}B_l(t) \tag{2.4.34}$$

Since the Hamiltonian function H is a function of Q and P, by using the Itô differential rule, Itô stochastic differential equations for $H(t)$ can be derived from (2.4.34) as follows:

$$\begin{aligned} \mathrm{d}H &= \sum_{j=1}^{n}\frac{\partial H}{\partial Q_j}\mathrm{d}Q_j + \sum_{j=1}^{n}\frac{\partial H}{\partial P_j}\mathrm{d}P_j + \frac{1}{2}\sum_{j,k=1}^{n}\frac{\partial^2 H}{\partial P_j\partial P_k}(\mathrm{d}P_j)(\mathrm{d}P_k) \\ &= \varepsilon\left(-\sum_{j,k=1}^{n} m_{jk}\frac{\partial H}{\partial P_j}\frac{\partial H}{\partial P_k} + \sum_{j,k=1}^{n}\sum_{j,s=1}^{m}\pi K_{ls}g_{jl}g_{ks}\frac{\partial^2 H}{\partial P_j\partial P_k}\right)\mathrm{d}t \\ &\quad + \varepsilon^{1/2}\sum_{j=1}^{n}\sum_{l=1}^{m} g_{jl}\frac{\partial H}{\partial P_j}\mathrm{d}B_l(t) \end{aligned} \tag{2.4.35}$$

The right of (2.4.35) is small, so the Hamiltonian $H(t)$ is a slow-changing process. Replacing P_1 with H, the system equation consists of Eqs. (2.4.34) and (2.4.35) except for the equation P_1. As H is a slow-changing process and

Q_1, Q_2, …, Q_n, P_2, …, P_n are fast-changing process, the stochastic averaging principle of Khasminskii can be applied. Futhermore, the Markov diffusion process approximates H, which is governed by the following Itô stochastic differential equations.

$$\mathrm{d}H = m(H)\mathrm{d}t + \sigma(H)\mathrm{d}B(t) \tag{2.4.36}$$

where $B(t)$ is a unit Wiener process, and the drift coefficient $m(H)$ and diffusion coefficient $\sigma(H)$ are derived from the time average of (2.4.35).

$$m(H) = \varepsilon\left\langle -\sum_{j,k=1}^{n} m_{jk}\frac{\partial H}{\partial P_j}\frac{\partial H}{\partial P_k} + \sum_{j,k=1}^{n}\sum_{j,s=1}^{m} \pi K_{ls}g_{jl}g_{ks}\frac{\partial^2 H}{\partial P_j \partial P_k}\right\rangle_t \tag{2.4.37}$$

$$\sigma^2(H) = \varepsilon^{1/2}2\pi\left\langle \sum_{j,k=1}^{n}\sum_{l,s=1}^{m} K_{ls}g_{jl}g_{ks}\frac{\partial H}{\partial P_j}\frac{\partial H}{\partial P_k}\right\rangle_t \tag{2.4.38}$$

It is noted that the state space response of the non-integrable Hamiltonian is ergodic on the energy surface of the $(2n - 1)$ dimension. The time averaging can be replaced by a space average of the fast-changing variables Q_1, Q_2, …, Q_n, P_2, …, P_n. Specifically, the time average of a function can be calculated as follows:

$$\langle F(\mathbf{q},\mathbf{p})\rangle_t = \frac{1}{T(H)}\int_{\Omega} F(\mathbf{q},\mathbf{p})\left(\frac{\partial H}{\partial p_1}\right)^{-1} \mathrm{d}q_1\mathrm{d}q_2\ldots\mathrm{d}q_n\,\mathrm{d}p_2\ldots\mathrm{d}p_n \tag{2.4.39}$$

where

$$T(H) = \int_{\Omega}\left(\frac{\partial H}{\partial p_1}\right)^{-1} \mathrm{d}q_1\mathrm{d}q_2\ldots\mathrm{d}q_n\,\mathrm{d}p_2\ldots\mathrm{d}p_n \tag{2.4.40}$$

While the integral domain Ω is defined as $H(q_1, q_2, \ldots, q_n, p_1 = 0, p_2, \ldots, p_n) \le H$.

After the stochastic average, the drift coefficient $m(H)$ and the diffusion coefficient $\sigma(H)$ are obtained, and the stationary-state probability density $p(h)$ of the Hamiltonian H is

$$p(h) = \frac{C}{\sigma^2(h)}\exp\left[\int \frac{2m(h)}{\sigma^2(h)}\mathrm{d}h\right] \tag{2.4.41}$$

References

1. Zhu, Weiqiu and Guoqiang Cai. 2017. *Introduction to Stochastic Dynamics*. Beijing: Science Press. (in Chinese).
2. Zhu, Weiqiu. 1992. *Random Vibration*. Beijing: Science Press. (in Chinese).
3. Zhu, Weiqiu. 2003. *Nonlinear Stochastic Dynamics and Control—Hamiltonian Theoretical Framework*. Beijing: Science Press. (in Chinese).
4. Mao, Xuerong. 2007. *Stochastic Differential Equations and Applications*. Cambridge: Elsevier.
5. Lin, Y.K. and Guoqiang Cai. 1995. *Probabilistic Structural Dynamics: Advanced Theory and Applications*. New York: McGraw-Hill.

Chapter 3
Stochastic Dynamic Simulation of Power System

Abstract Constructing or selecting an appropriate model is the foundation for the study of stochastic dynamics. In previous power system analysis, the disturbances to the system are mostly described by deterministic models, such as step-type models and periodic models . And deterministic time-domain or frequency-domain models are often used for system analysis. When studying the dynamic behavior of power systems under stochastic disturbances, the model must be established based on the stochastic disturbance characteristics in the power system, and it is necessary to consider how to introduce the stochastic disturbance model into the construction of the power system model. In this chapter, the ideal and measured stochastic disturbance models are presented, and a stochastic model of power system is constructed to analyze and compare the stochastic dynamic response calculation methods.

3.1 Introduction

The simulation is an indispensable means in power system dynamic analysis and control [1–7], and it is also the essential basis for power system operators to guide the safe operation of the power grid [8–12]. Because the simulation calculations are based on models, inappropriate models can lead to inconsistencies in calculation results with the actual situation, either optimistic or conservative [13–18]. Under critical conditions, they may even change qualitative conclusions or cover up some important features that may constitute potential danger [19–24]. For studying on power system oscillation under stochastic disturbance, it is very important to construct or select an appropriate model, which is an important guarantee for the follow-up mechanism analysis and validation of the example [25–29].

In previous power system analysis, the disturbances to the system are mostly described by deterministic models, such as step-type models and periodic models, and deterministic time-domain or frequency-domain models are often used for system analysis [30–35]. When studying the dynamic behavior of power systems under stochastic disturbances, the stochastic disturbance characteristics in the power

P. Ju, *Stochastic Dynamics of Power Systems*, Power Systems,
https://doi.org/10.1007/978-981-13-1816-0_3

system should be involved, and it is necessary to consider how to introduce the stochastic disturbance model into the construction of the power system model [36–42].

3.2 Stochastic Disturbance Model of Power System

3.2.1 Ideal Stochastic Disturbance Model

The stochastic disturbance in the power system can be described as a stochastic process, and a stochastic process can be modeled as a series of stochastic variables at different moments. For stochastic processes, classification can be made according to different characteristics. According to their probability distributions, the stochastic processes can be divided into Gaussian processes, Rayleigh processes, Poisson processes and so on. According to the bandwidth of spectral density, the stochastic processes can be divided into narrowband processes, wideband processes, and white noises of infinite bandwidth and so on. According to their evolutionary nature over time, the stochastic processes can be divided into smooth and non-stationary processes. White noise refers to the noise where a power spectral density (PSD) that exhibits a uniform distribution over the entire frequency domain, that is, it has the same energy density at all frequencies. In fact, such kind of ideal stochastic signal with infinite bandwidth and unlimited energy does not exist, but it is easier for mathematical processing and has been widely used in the fields of electronics, communications, finance and architecture. It is generally considered that if a noise process has a power spectrum width much larger than the bandwidth of the system in which it is applied, and its power spectral density in the bandwidth can be regarded as a constant, such noise process can be treated as a white noise, such as the thermal noise common in an actual project.

A mathematically defined stationary stochastic process with a non-zero constant power spectrum density is called a white noise process, abbreviated as a white noise. If the white noise probability density function satisfies the normal distribution characteristic, it is called Gaussian white noise. This chapter defines the Gaussian white noise process as $W(t),\ t > 0$. It has the same energy density in the entire frequency domain and has the following 3 properties:

(1) The mathematical expectation is 0, i.e., $E[W(t)] = 0$;
(2) The autocorrelation function is $R(\tau) = N_0\delta(\tau)$, where N_0 is a constant and $\delta(\tau)$ is a Dirac function;
(3) The power spectral density is constant, i.e., $S(\omega) = N_0$ and $-\infty < \omega < +\infty$.

Since the stochastic factors in the power system have the characteristics of wide distribution and variety, it is often difficult to describe their stochastic characteristics uniformly when considering the impacts of stochastic disturbances on power system dynamic characteristics. Therefore, when constructing the stochastic disturbance

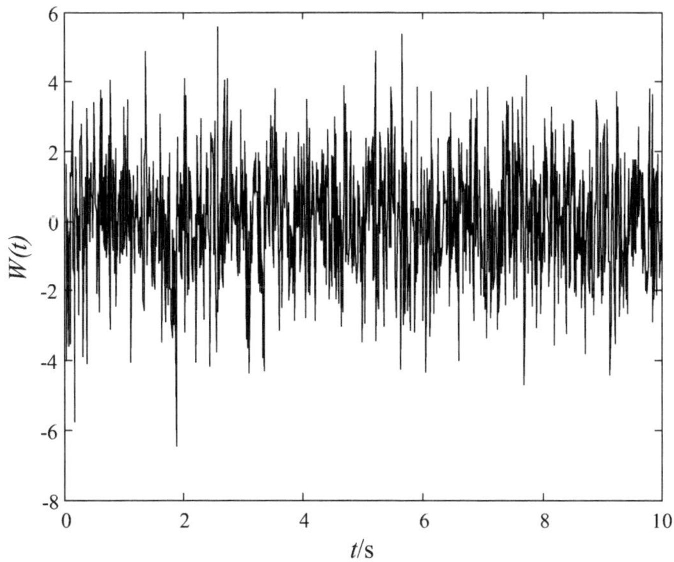

Fig. 3.1 Gaussian white noise process

model of the power system, it can be assumed that the stochastic disturbance is an ideal Gaussian white noise process. This is for the convenience of analysis. In addition, it is also because that the theory based on the ideal distribution has made relatively great progresses, such as the theory of stochastic differential equations, etc., which can provide a series of effective analytical tools.

According to the mathematical definition of the Gaussian white noise, it may be set to a finite length signal, and use Matlab to simulate its discrete form. The total simulation time is $T = 10$ s, the simulation time step length is $\Delta t = 0.01$ s, and the Gaussian white noise discrete point number is $N = 1000$, with the result shown in Fig. 3.1.

In order to ensure the simulation of Gaussian white noise in Matlab effectiveness, the autocorrelation function and the power spectral density of the simulation results can be verified according to the nature of Gaussian white noise. First, the autocorrelation function of the Gaussian white noise discrete sequence is calculated by using Eq. (3.2.1):

$$R_{xx}(m) = \frac{1}{N}\sum_{n=1}^{N-m} x(n)x(n+m), \quad m = 0, 1, \ldots, N-1 \tag{3.2.1}$$

Second, the Wiener-Khintchine theorem is used to calculate the inverse Fourier transform of $R_{xx}(m)$, and the calculation formula is as follows:

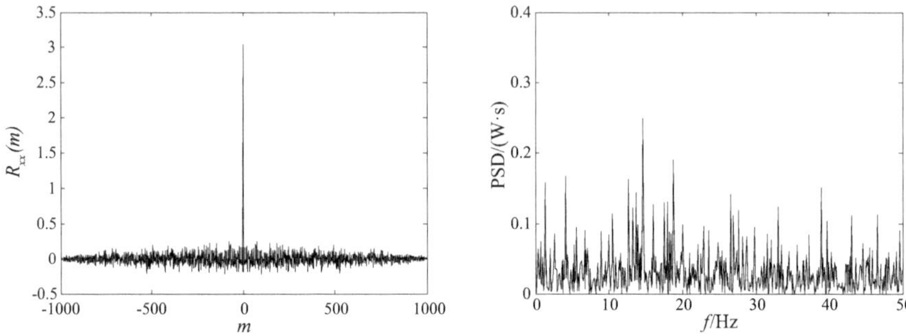

(a) Autocorrelation function **(b) Power spectral density**

Fig. 3.2 Autocorrelation function and power spectrum characteristics of Gaussian white noise process

$$S(\omega) = \sum_{m=0}^{N-1} R_{xx}(m)\mathrm{e}^{-j\omega m} \tag{3.2.2}$$

The autocorrelation function and the power spectral density of the results of Gaussian white noise simulation can be obtained from the above formulas, which are shown in Fig. 3.2.

It can be seen that the autocorrelation function of Gaussian white noise has the property of Dirac function, that is, the function value is 0 except the zero point, and the power spectral density function of Gaussian white noise is approximately constant at $(-\infty, +\infty)$, and is uniformly distributed throughout the entire frequency domain. It can be concluded that the simulation results fully conform to the properties of Gaussian white noise described in the mathematical definition.

3.2.2 Measured Stochastic Disturbance Model

From the point of view of practical applications, there are some limitations to describing the stochastic disturbance as the ideal Gaussian white noise model. The stochastic disturbance in the power system is often not an ideal Gaussian distribution or white noise process, so naturally it is desirable to describe the stochastic disturbance using its actual characteristics. This requires investigating the statistical information of the measured data, or performing reasonable assumptions by combining facts. This section also made some preliminary attempts in modeling stochastic disturbance based on measured data. With the currently collected power fluctuation data such as wind power and photovoltaic, which are apparently stochastic, the time domain fluctuation curve appears to be dazzling and confused, and cannot be analyzed to draw valid conclusions. However, by converting these stochastic disturbance data from the time domain to the frequency domain, it is found that the frequency domain can more clearly describe the laws and characteristics of stochastic disturbances, and there are certain similarities in the frequency domain characteristics of stochastic disturbance data at different time periods. Therefore, when investigating stochastic disturbance modeling, the frequency domain model is better to describe the inherent characteristics of stochastic disturbances. This section takes the actually measured wind farm active power fluctuation data as an example, analyzes its frequency domain characteristics from the perspective of frequency domain, and tries establishing a frequency domain model for it, aiming to provide a new way to analyze the characteristics of stochastic disturbance in power systems, and to model for the frequency domain.

The frequency domain analysis technology is usually based on a continuous Fourier transform, and the frequency domain characteristics of the sampled signals are studied by using both the amplitude spectrum and the phase spectrum. When the deterministic signal is analyzed, the conversion results of any samples are the same or similar in case that the sampling frequency and the number of sampling points are determined. But for stochastic process signals, the Fourier transform does not exist in a strictly mathematical sense. In the actual processing, when making stochastic sampling upon stochastic signals, the conversion results are also different from each other, and are difficult to analyze. The stochastic disturbance in the power system is essentially a stochastic signal that cannot be analyzed in the frequency domain by a continuous Fourier transform.

To solve the problem of frequency domain analysis of stochastic process signals, for the study of stationary stochastic processes, the power spectral density function PSD is introduced, and its expression is as follows:

$$S_x(f) = \lim_{T\to\infty} \frac{1}{T} E[|X(f,T)|^2] \tag{3.2.3}$$

The mean value function $E[\cdot]$ is to average the Fourier transform results of each sample of the stochastic signal, and then average the energy spectral density $|X(f,T)|^2$ of the signal on the cutoff time to obtain the power distribution of the signal at each frequency point, and approach the actual frequency domain characteristics of stochastic signals. Therefore, the power spectrum can be used to describe the distribution of power fluctuations in the stationary stochastic process in the frequency domain.

The definition of the power spectrum of a stationary stochastic process is based on an infinite and continuous system, in order to facilitate the strict and precise description of the frequency domain characteristics of the stochastic process. However, in practical applications, due to the limitation of the computer's computing power, it is necessary to derive the discrete calculation formula for the power spectrum of a stationary stochastic process.

For a stochastic process $x(t)$ with a length of T and a time domain sampling interval of T_s, the time domain sampling points may be represented as $t_k = (k-1)T_s$ and $k = 1, 2, \ldots, N$. The symmetric extension $x(t)$ is:

$$x_T(t) = \begin{cases} x(t) & 0 \le t \le T \\ x(-t) & -T \le t \le 0 \\ 0 & |t| > T \end{cases} \tag{3.2.4}$$

According to Eq. (3.2.3), the sampling of the power spectral density of the stochastic process $x(t)$ in the frequency domain can be written as:

$$S_x(f_i) = \lim_{T \to \infty} \frac{1}{2T} E(|X(f_i, T)|^2) \tag{3.2.5}$$

By defining the sampling frequency as $f_s = 1/T_s$ and the frequency domain resolution as $\Delta f = f_s/N$, the frequency domain sampling points in Eq. (3.2.5) can be expressed as $f_i = \pm(i-1)\Delta f$ and $i = 1, 2, \ldots, N$. First, calculate $X(f_i, T)$, and according to the symmetry of $x_T(t)$, the result of sampling the continuous Fourier transform in the frequency domain is:

$$X(f_i, T) = \int_{-T}^{T} x_T(t)\mathrm{e}^{-j2\pi f_i t}\mathrm{d}t = 2\int_{0}^{T} x(t)\mathrm{e}^{-j2\pi f_i t}\mathrm{d}t \tag{3.2.6}$$

Set $\mathrm{d}t \to \Delta T = T_s$, and use Riemann integral expression, the above equation can be expressed as:

$$X(f_i, T) = 2\sum_{k=1}^{N} x(t_k)\mathrm{e}^{-j2\pi f_i t_k} T_s \tag{3.2.7}$$

Substituting into the frequency domain sampling point f_i and the time domain sampling point t_k expression, we can get:

$$X(f_i, T) = 2T_s \sum_{k=1}^{N} x(t_k)\mathrm{e}^{-j2\pi(i-1)\Delta f\cdot(k-1)T_s} = 2T_s \sum_{k=1}^{N} x(k)\mathrm{e}^{-j2\pi\frac{(i-1)(k-1)}{N}} \tag{3.2.8}$$

The summation in Eq. (3.2.8) can be calculated using the fast Fourier transform function $FFT(\cdot)$ in Matlab. Therefore, the sampling of the power spectrum of the stochastic process $x(t)$ shown in Eq. (3.2.5) in the frequency domain can be approximated as:

$$\begin{aligned} S_x(f_i) &= \frac{1}{2T} E(|2T_s \cdot FFT(x(k))|^2) \\ &= \frac{1}{2NT_s} \cdot 4T_s^2 E(|FFT(x(k))|^2) \\ &= \frac{2E(|FFT(x(k))|^2)}{Nf_s} \end{aligned} \tag{3.2.9}$$

It should be noted that the above equation is the derivation based on the bilateral spectrum. In the actual project, since the analysis is performed only in the positive frequency range, and the commonly used unilateral spectrum can be expressed as $S'_x(f_i)$, the expression is shown in (3.2.10):

$$S'_x(f_i) = 2S_x(f_i) - \frac{E(|2FFT(x(k))|^2)}{Nf_s}, \quad f_i \geq 0 \tag{3.2.10}$$

For a stationary stochastic process $x(t)$ with n samples $(x_1(t), x_2(t), \ldots x_n(t))$, the discrete calculation formula of power spectral density is:

$$\begin{aligned} S'_x(f_i) &= \frac{E\left(|2FFT(x(k))|^2\right)}{Nf_s} \\ &= \frac{E\left(|\frac{2}{N}FFT(x(k))|^2\right)}{\Delta f} \\ &= \frac{\frac{1}{n}\sum_{j=1}^{n}|\frac{2}{N}FFT(x(k))|^2}{\Delta f}, \quad f_i \geq 0 \end{aligned} \tag{3.2.11}$$

In such equation, $E\left(|\frac{2}{N}FFT(x(k))|^2\right)$ represents the power of the sampling point of each frequency; Δf is the frequency resolution, which means the sampling interval in the frequency domain. Dividing the two to produce the power spectral density of the sampling signal, its unit is W s.

Based on the power spectrum calculation formula for stochastic signals shown in Eq. (3.2.11), frequency domain analysis of wind power stochastic power fluctuations can be performed. The sampling data is the active power data of a wind farm on January 1, 2015, and the sampling interval is 0.02 s. Affected by the natural characteristics of the wind level, the active power output of the wind farm has obvious peaks and troughs in one day. In general, the peak period occurs in the morning and evening of the day, and the trough period appears at noon. Therefore, the sampling data of the peak, trough and medium output of the wind farm are selected as samples respectively, and the time and frequency domain characteristics of wind power under different output levels are compared and analyzed.

Figure 3.3a shows the active power fluctuation curve of the wind farm from 4 to 5 a.m. At this time, the active power of the wind farm is between 38 and 50 MW, which is at the peak output level. It can be seen that the active power of the wind farm changes with a slow fluctuation as a whole, but at the same time, it is accompanied by a relatively small and rapidly varying power fluctuation. Similar conclusions can also be drawn from the results of its power spectrum analysis. The energy of wind power fluctuations is mainly concentrated in the low frequency band and the ultra-low frequency band. The energy in the middle and high frequency bands is almost 0. The lower the frequency is, the larger the amplitude of the corresponding power spectrum is. Due to the large differences in the power spectrum amplitudes among different frequency bands (e.g., the power spectrum amplitude at 0.0002 Hz is 29,730, the amplitude at 0.002 Hz is 142, and the amplitude at 0.02 Hz is only 4.8), and if all bands are plotted in the power spectrum, it is impossible to visually observe the change trend of wind power in the frequency domain. Therefore, one frequency band of the power spectrum is plotted to represent the general variation of wind power fluctuations in the frequency domain. For example, if we focus on the impact of wind power fluctuations on power system oscillations, we can analyze the power spectrum of wind power fluctuations in the low frequency band (0.1–2.5 Hz), which is shown in Fig. 3.3b.

Similarly, Figs. 3.4 and 3.5 show the time-domain fluctuation curve of active power and its corresponding power spectrum when wind farm output is low (6–18 MW) and medium (20–35 MW), respectively. It can be seen that there exist relatively large differences in the time-domain fluctuation curves of wind power at different output levels, and no valid conclusion can be obtained from this analysis. However, there is obvious similarity in the energy distribution of wind power

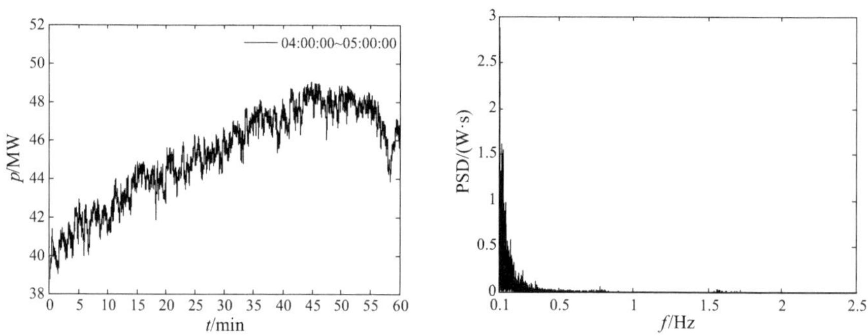

(a) Time-domain curve **(b) Power spectrum**

Fig. 3.3 Time-domain fluctuation curve and power spectrum of wind farm peak output

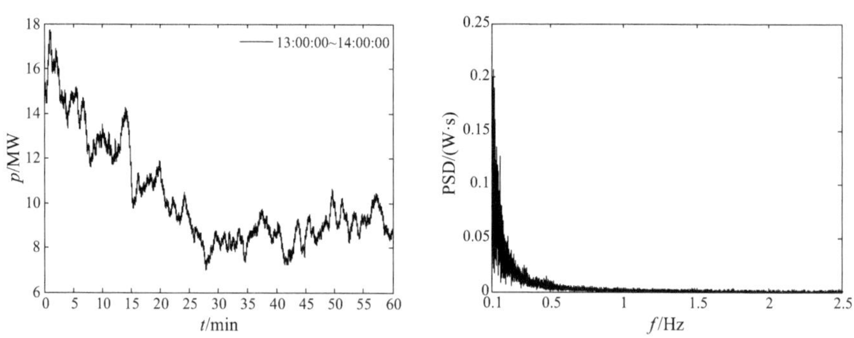

(a) Time-domain curve **(b) Power spectrum**

Fig. 3.4 Time-domain fluctuation curve and power spectrum of wind farm trough output

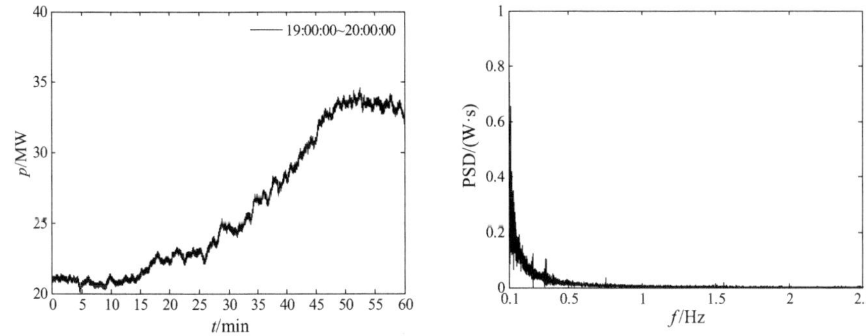

(a) Time-domain curve **(b) Power spectrum**

Fig. 3.5 Time-domain fluctuation curve and power spectrum of wind farm medium output

fluctuations in the frequency domain, and only the power spectrum amplitude is different due to the difference in the output level of the wind farm. At the same time, compared to the time domain characteristics, the trend of the frequency domain characteristics of wind power is more regular, and its power spectrum amplitude decreases with the increase of frequency. Based on this conclusion, we can describe wind power accurately in the frequency domain and establish a suitable frequency domain model.

It can be concluded from the above analysis that the amplitudes of the power spectrum of wind power fluctuations in different frequency bands are quite different. When attempting to construct a corresponding function model to directly fit the wind power spectrum, the power spectrum data points in the low-frequency band and the middle-high frequency band are often ignored, and only the large-amplitude power spectrum data in the ultra-low frequency band are fitted. This phenomenon causes the overall fitting effect to be unsatisfactory although the calculated fitting error is not large. Therefore, in order to weaken the amplitude difference of wind power spectrum in different frequency bands, the logarithm of its amplitude and frequency are taken into consideration, and the power spectrum is converted into logarithmic spectrum. For example, the wind power spectrum of the peak output shown in Fig. 3.3 converted to the logarithmic spectrum is shown in Fig. 3.6. It can be seen that the fluctuation of wind power has a similar linear variation in the logarithmic coordinate system. Therefore, in the logarithmic coordinate system, the linear function model $f(x) = a + bx$ is used to fit the change law.

The method chosen for the fitting process is the weighted least square method. The requirement and the principle for curve fitting using this method are: for a given set of sample data $(x_i, y_i)(i = 1, 2, \ldots, m)$, find a function $f(x)$ such that the sum of squared weighted deviations satisfies:

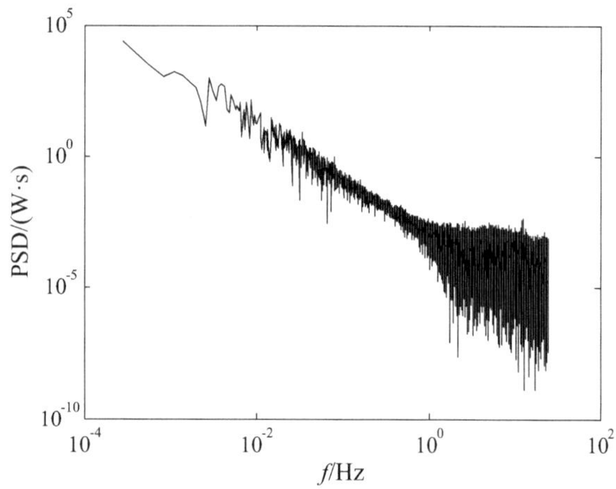

Fig. 3.6 Logarithmic spectrum of wind power peak output

$$\sum_{i=1}^{m} w_i[f(x_i) - y_i]^2 = \min \sum_{i=1}^{m} w_i[f(x_i) - y_i]^2 \tag{3.2.12}$$

where w_i is the weight coefficient of the sample data $(x_i,\ y_i)$, it represents the contribution of the sample data points to the curve fitting effect. All weight coefficients are positive and satisfy $\sum_{i=1}^{m} w_i = 1$. For data with higher accuracy or more important status, a relatively greater weight should be given. For the above selected linear fitting function $f(x) = a + bx$, the solution to Eq. (3.2.12) can be reduced to solve Eq. (3.2.13) according to the principle of the weighted least square method.

$$\begin{bmatrix} \sum_{i=1}^{m} w_i & \sum_{i=1}^{m} w_i x_i \\ \sum_{i=1}^{m} w_i x_i & \sum_{i=1}^{m} w_i x_i^2 \end{bmatrix} \begin{bmatrix} a \\ b \end{bmatrix} = \begin{bmatrix} \sum_{i=1}^{m} w_i y_i \\ \sum_{i=1}^{m} w_i x_i y_i \end{bmatrix} \tag{3.2.13}$$

It should be noted that the logarithmic operations performed on the raw wind power spectrum data actually change the raw data of equal degrees of certainty into unequal degrees of certainty. When such uncertainty is within a reasonable range, it has a small effect on the fitting effect; but when the uncertainty is beyond the reasonable range, the fitting effect will be affected. In simple terms, the data points of the original wind power spectrum data are equal in an arbitrary interval scale range, but after converting to a logarithmic spectrum, there is relatively large gaps between data points in the arbitrary interval scale range. For example, there is a 10-fold relationship between data points in the range of 11–10 Hz and 10–100 Hz, and both of the ranges have the same length. This means that each power spectral data point in the low-frequency part contributes more to the curve fitting than that in the high-frequency part, thereby affecting the curve fitting effect. In order to correct the unequal degree of certainty of each power spectrum data point, the weight coefficient is set as:

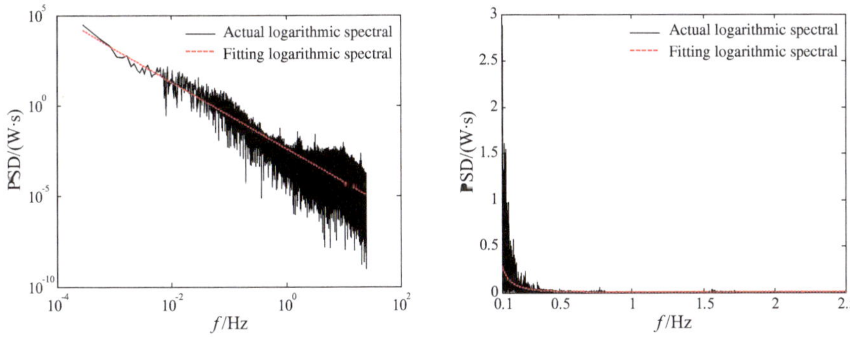

(a) Logarithmic spectral fitting result **(b) Restored power spectrum**

Fig. 3.7 Wind power peak output logarithmic spectrum fitting result and restored power spectrum

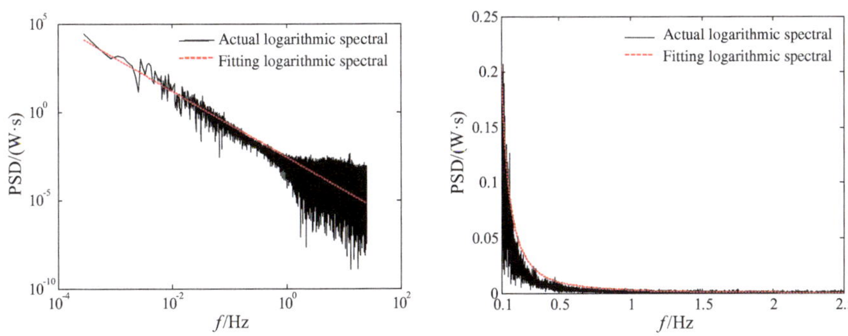

(a) Logarithmic spectral fitting result **(b) Restored power spectrum**

Fig. 3.8 Wind power trough output logarithmic spectrum fitting result and restored power spectrum

$$w_i = \frac{x_{i+1} - x_i}{x_m - x_1}, \quad i = 1, 2, \ldots, m - 1 \tag{3.2.14}$$

After the appropriate weight coefficients are set, the logarithmic spectrum shown in Fig. 3.6 is fitted using this weighted least square method, and the fitting results are restored to the power spectrum. The results of the fitting areas are shown in Fig. 3.7.

Similarly, when the logarithmic spectrum of wind farms with low and medium output is fitted, the results of the fitting are shown in Figs. 3.8 and 3.9. It can be seen that the weighted least squares method can achieve good results for the fitting of different logarithmic spectrum sample data; and after the logarithmic spectrum fitting result is restored to the power spectrum, the fitting curve obtained is also more suitable for the original wind power spectrum data. At the same time, compared with

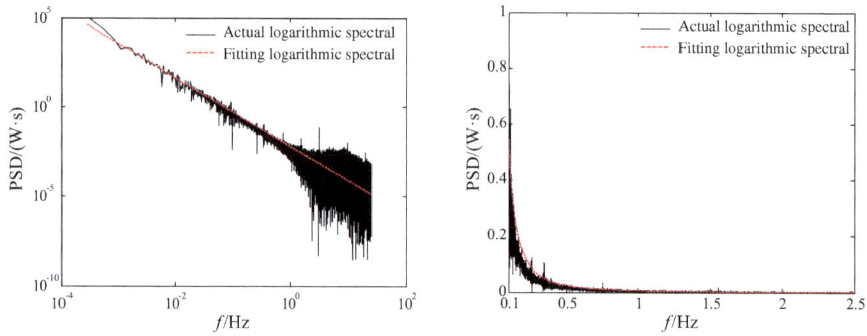

(a) Logarithmic spectral fitting result **(b) Restored powerspectrum**

Fig. 3.9 Wind power middle output logarithmic spectrum fitting result and restored power spectrum

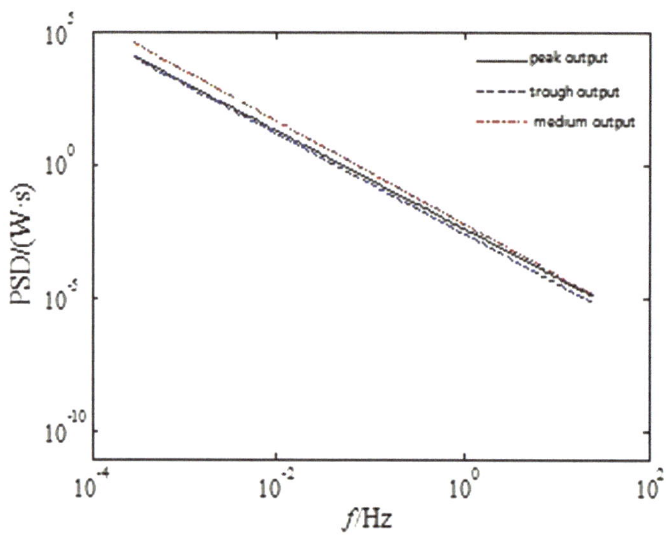

Fig. 3.10 Comparison of logarithmic spectrum fitting results for different output of wind farm

the results of fitting directly to the wind power spectrum, the fitting results after the transformation of logarithmic spectrum can better represent the trend of wind power in the low-frequency band and the middle-high frequency band.

In addition, Fig. 3.10 shows the comparison of the results of the logarithmic spectrum fitting under different outputs of the wind farm. It can be seen that the fitting lines of the wind power logarithmic spectrum are close under different output conditions. This phenomenon further verifies the similarity of the frequency domain characteristics of wind power, and also shows that the established frequency domain model has a certain universality. This modeling idea can be extended to other stochastic disturbance modeling studies in power systems.

3.3 Stochastic System Model of Power System

3.3.1 Stochastic Model of Single-Machine Infinite Bus System

The single-machine infinite bus system is shown in Fig. 3.11. Based on this, basic theoretical problems such as the electromechanical transient process characteristics and its stability under stochastic disturbances are studied. The generator uses a second-order swing equation, and we assume that: (1) the internal potential E' is constant; (2) the transient salient-pole effect is ignored; (3) the mechanical power P_m is constant.

The stochastic disturbances can be considered as stochastic power fluctuations generated by loads such as renewable energy generators or electric vehicles, which result in imbalance between mechanical power and electromagnetic power. The stochastic disturbance term P_L is used to represent the sum of these power fluctuations, so P_L can be a positive or negative value. The stochastic model of the electromechanical transient process for a single-machine infinite bus system is:

$$M\frac{\mathrm{d}^2\delta}{\mathrm{d}t^2} + D\frac{\mathrm{d}\delta}{\mathrm{d}t} = P_m - P_{\max}\sin(\delta - v) - P_L \tag{3.3.1}$$

where

$$P_L = \sigma W(t) \tag{3.3.2}$$

Each variable in the equation is the per-unit value, δ—rotor angle, M—inertial time constant, D—damping coefficient, t—time, $P_{\max}$—maximum electromagnetic power, v—impedance angle, $W(t)$—standard Gaussian process, σ^2—variance of P_L.

3.3.2 Stochastic Model of Two-Machine Power System

The two-machine power system is used to study the effect of a stochastic disturbance in the interval oscillation mode of the interconnected power grid, and the

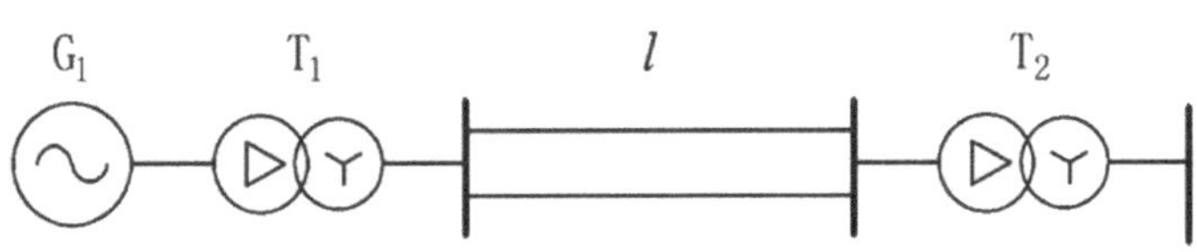

Fig. 3.11 Single-machine infinite bus system

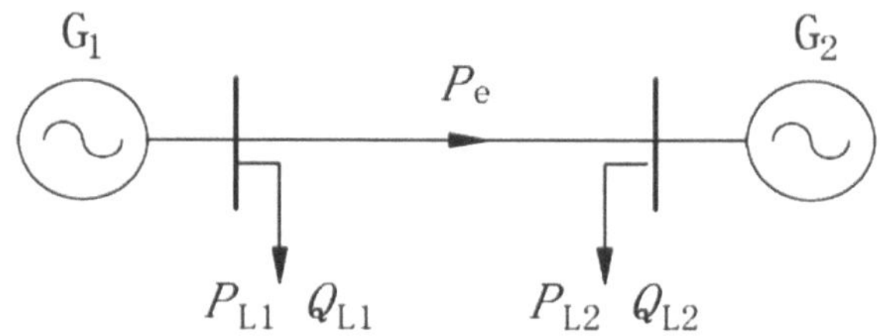

Fig. 3.12 Two-machine system

two-machine system is shown in Fig. 3.12. Under a stochastic disturbance, the two generators swing equations are:

$$\begin{cases} M_1 \dfrac{d^2\delta_1}{dt^2} + D_1 \dfrac{d\delta_1}{dt} = P_{m1} - P_e - P_{L1} \\ M_2 \dfrac{d^2\delta_2}{dt^2} + D_2 \dfrac{d\delta_2}{dt} = P_{m2} + P_e - P_{L2} \end{cases} \tag{3.3.3}$$

The model is simplified and processed by introducing the hypothesis:

$$M\frac{d^2\delta}{dt^2} + D\frac{d\delta}{dt} = P_{\max}\sin(\delta_0 - v) - P_{\max}\sin(\delta - v) - \sigma W(t) \tag{3.3.4}$$

where $\delta = \delta_1 - \delta_2$, $M = \frac{M_1 M_2}{M_1 + M_2}$, $D = \beta_2 D_1 = \beta_1 D_2$ and $\sigma = \sqrt{\beta_2^2\sigma_1^2 + \beta_1^2\sigma_2^2}$.

3.3.3 Stochastic Model of Multi-machine Power System

The dynamic process and its characteristics caused by the stochastic disturbance are studied when there are multiple weakly damped modes in a multi-machine system. A multi-machine power system is a high-order and nonlinear system, and the model is more complex after adding stochasticity, so some simplifying assumptions are made. For the *i*-th generator node, the stochastic fluctuation of its mechanical power is represented in P_{mi}; for the power fluctuation of renewable energy generators, it can be understood as a node injection power. Therefore, the above stochastic disturbance can be expressed approximately as a superposition of multiple different stochastic processes:

$$P_{Li} = \sum_k \sigma_{ik} W_{ik}(t) \tag{3.3.5}$$

In such equation: $W_{ik}(t)$ and σ_{ik}^2 represent the standard Gaussian process and the variance of different frequencies, which are independent of each other. High-frequency components generally correspond to stochastic power fluctuations caused by electrical factors such as electric vehicles, and low-frequency components generally correspond to stochastic power fluctuations caused by mechanical factors such as wind power generators.

The corresponding stochastic differential equation is:

$$M_i \frac{d^2 \delta_i}{dt^2} + D_i \frac{d\delta_i}{dt} = P_{mi} - P_{ei} - P_{Li} \tag{3.3.6}$$

3.3.4 Regular Form of Power System Stochastic Model

The Itô-type stochastic differential equation is used here, and its general expression can be written as:

$$\mathrm{d}\boldsymbol{X}(t) = f(\boldsymbol{X}(t), t)\mathrm{d}t + g(\boldsymbol{X}(t), t)\mathrm{d}\boldsymbol{B}(t) \tag{3.3.7}$$

where $\boldsymbol{X}(t) = (X_1(t), X_2(t), \ldots, X_n(t))^{\mathrm{T}}$ and $\boldsymbol{B}(t) = (B_1(t), B_2(t), \ldots, B_n(t))^{\mathrm{T}}$ represent the n-dimensional vector stochastic variable and the n-dimensional Brownian motion, respectively, and the initial values $\boldsymbol{X}(t_0)$ and $\boldsymbol{B}(t)$ are independent.

In 1827, R. Brown observed the irregular physical movement of pollen in an aqueous solution of pollen particles, and such movement is known as the Brownian motion. Subsequently, in 1918, Wiener proposed a strict mathematical model describing Brownian motion using stochastic process language. Therefore, Brownian motion can also be called the Wiener process, and its mathematical definition is described in 3.1. The study of Brownian motion marks the beginning of the study on stochastic differential equations.

Definition 3.1 The standard Brownian motion or the standard Wiener process defined on $[0, T]$ is a stochastic process $B(t)$ that satisfies the following three conditions:

(1) $B(0) = 0$;
(2) For $0 \le s < t \le T$, increment $B(t) - B(s) \sim \sqrt{t - s}N(0, 1)$;
(3) For $0 \le s < t < u < v \le T$, increments $B(t) - B(s)$ and $B(u) - B(v)$ are independent.

Studies have shown that the Brownian motion process $B(t)$ has the following properties:

Theorem 3.1 *For any different $t, s > 0$, the related function about $B(t)$ and $B(s)$ satisfies:*

$$R_B(t, s) = E[B(t)B(s)] = \min(t, s) \tag{3.3.8}$$

Proof When $t > s > 0$,

$$\begin{aligned} E[B(t)B(s)] &= E\{[B(t) - B(s)]B(s) + B(s)B(s)\} \\ &= E\{[B(t) - B(s)]B(s)\} + E[B(s)B(s)] \\ &= E\{[B(s)]^2\} = s \end{aligned}$$

Similarly, when $s > t > 0$, $E[B(t)B(s)] = t$

That is, $E[B(t)B(s)] = \min(t, s)$ makes sense.

According to Theorem 3.1, we can get the following relationship between the Brownian motion process and Gaussian white noise process:

Theorem 3.2 *Assuming that the formal derivative of the Brownian motion* $B(t)$ *is* $\dot{B}(t)$, $\dot{B}(t)$ *has the property of a Gaussian white noise process.*

Proof

(1) With the property of Brownian motion, we can obtain $E[\dot{B}(t)] = 0$;
(2) According to Theorem 3.1, we can obtain by $R_B(t, s) = \min(t, s)$

$$R_{\dot{B}}(t, s) = E[\dot{B}(t)\dot{B}(s)] = \frac{\partial^2 R_B(t, s)}{\partial t \partial s} = \frac{\partial u(t - s)}{\partial t} = \delta(t - s) \tag{3.3.9}$$

where for a given s, the function $u(t - s)$ is a Heaviside unit step function of t, i.e.,

$$u(t - s) = \begin{cases} 0, & t < s; \\ 1, & t > s. \end{cases}$$

$\delta(\cdot)$ represents the δ-function, and in the above equation, when $t \neq s$, $\dot{B}(t)$ is not related to $\dot{B}(s)$, that is, the correlation function of $\dot{B}(t)$ is $R_{\dot{B}}(t, s) = \delta(t - s)$;

(3) According to the Wiener-Khintchine theorem, the power spectrum of $\dot{B}(t)$ can be calculated to be $S_{\dot{B}}(\omega) = 1, \ -\infty < \omega < +\infty$.

In summary, the formal derivative $\dot{B}(t)$ of the Brownian motion process $B(t)$ is characterized by Gaussian white noise. According to Theorem 3.2 and the definition of the Gaussian white noise process, the relationship between the two can be expressed as follows:

$$\frac{\mathrm{d}B(t)}{\mathrm{d}t} = W(t) \tag{3.3.10}$$

By combining Eqs. (3.3.1) and (3.3.4), we can find that the Itô-type stochastic differential equation used in this section is a type of dynamic equation driven by

Gaussian white noise, which can be extended to the study on the dynamic behavior of power systems under stochastic disturbances.

Take the single-machine infinite bus system as an example and let $X_1 = \delta$ and $X_2 = \frac{\mathrm{d}x_1}{\mathrm{d}t} = \frac{\mathrm{d}\delta}{\mathrm{d}t}$, a nonlinear vector stochastic differential equation model for the power system under ideal stochastic disturbance can be obtained by Eq. (3.3.1):

$$\mathbf{d}\boldsymbol{X} = \boldsymbol{F}(\boldsymbol{X})\mathrm{d}t + \boldsymbol{Q}(t)\mathrm{d}B(t) \tag{3.3.11}$$

where $\boldsymbol{X} = \begin{pmatrix} X_1 \\ X_2 \end{pmatrix} = \begin{pmatrix} \delta \\ \frac{\mathrm{d}\delta}{\mathrm{d}t} \end{pmatrix}$, $\boldsymbol{F}(\boldsymbol{X}) = \begin{pmatrix} X_2 \\ \frac{P_m}{M} - \frac{P_{\max}}{M}\sin X_1 - \frac{D}{M}X_2 \end{pmatrix}$ and $\boldsymbol{Q}(t) = \begin{pmatrix} 0 \\ \frac{\sigma_0}{M} \end{pmatrix}$.

Similarly, considering system damping and the linearized second-order model equation of the OMIB system, the stochastic variables representing stochastic disturbances are introduced as follows:

$$\begin{cases} M\dfrac{\mathrm{d}\Delta\omega}{\mathrm{d}t} = \Delta P_m - \Delta P_e + \sigma_0 W(t) \\ \dfrac{\mathrm{d}\Delta\delta}{\mathrm{d}t} = \Delta\omega \end{cases} \tag{3.3.12}$$

where $\Delta P_e = \left(\frac{EU}{X_\Sigma}\cos\delta_0\right)\Delta\delta$, and δ_0 is the generator rotor angle at the steady state operating point. After simplification, a systematic linear stochastic differential equation model with stochastic disturbance terms can be obtained:

$$\begin{cases} \dfrac{\mathrm{d}\Delta\delta}{\mathrm{d}t} = \Delta\omega \\ \dfrac{\mathrm{d}\Delta\omega}{\mathrm{d}t} = -\dfrac{E'U\cos\delta_0}{MX_\Sigma}\Delta\delta - \dfrac{D}{M}\Delta\omega + \dfrac{\sigma_0}{M}W(t) \end{cases} \tag{3.3.13}$$

Its vector form is:

$$\mathbf{d}\boldsymbol{X} = \boldsymbol{A}\boldsymbol{X}(t)\mathrm{d}t + \boldsymbol{Q}(t)\mathrm{d}B(t) \tag{3.3.14}$$

where $\mathbf{X}(t) = \begin{pmatrix} \Delta\delta \\ \Delta\omega \end{pmatrix}$, $\mathbf{X}(t_0) = \begin{pmatrix} 0 \\ 0 \end{pmatrix}$, $\boldsymbol{A} = \begin{pmatrix} 0 & 1 \\ a & b \end{pmatrix}$, $a = -\frac{E'U\cos\delta_0}{MX_\Sigma}$, $b = -\frac{D}{M}$ and $\boldsymbol{Q}(t) = \begin{pmatrix} 0 \\ \frac{\sigma_0}{M} \end{pmatrix}$.

3.4 Computation of Power System Stochastic Response

3.4.1 Numerical Computation Method

In general, it is difficult to derive the expression of the solution process for stochastic differential equations, and it is especially important to construct appropriate numerical methods to solve this problem. The Eular-Maruyama method is a common method for solving such equations. However, due to its low convergence order ($\gamma = 0.5$), Heun made further improvements and constructed a Heun numerical method with a strong convergence order of 1. This section is aimed at the calculation and the analysis of power system response characteristics under a stochastic power disturbance, in which it has higher requirements on the accuracy of the numerical solution process trajectory of the system model. Therefore, the Heun numerical method is more appropriate in the selection of numerical methods.

For a positive integer N against the stochastic differential equation of the form like Eq. (3.3.1), and let the stochastic process $\mathbf{X}(t)$ be its solution process, if:

$\Delta t = (T - t_0)/N$, $\tau_j = j\Delta t$, $\mathbf{X}_j = \mathbf{X}(\tau_j)$ and $j = 0, 1, 2, \ldots, N$

The differential iteration scheme of the Heun numerical method is:

$$\mathbf{X}_{j+1} = \mathbf{X}_j + \frac{f(X_j) + f(\hat{\mathbf{X}})}{2}\Delta t + \frac{g(X_j) + g(\hat{\mathbf{X}})}{2}\Delta B_j \tag{3.4.1}$$

where: $\hat{\mathbf{X}} = \mathbf{X}_j + f(X_j)\Delta t + g(X_j)\Delta B_j$.

According to Eq. (3.3.11), the simulation iteration scheme based on Heun numerical method for the stochastic differential equation model of OMIB system can be derived.

For a positive integer N against the system nonlinear stochastic differential equation model (3.3.7), if:

$\Delta t = (T - t_0)/N$, $\tau_j = j\Delta t$, $\omega_j = \omega(\tau_j)$, $\delta_j = \delta(\tau_j)$ and $j = 0, 1, 2, \ldots, N$

There are the following iteration formulas:

$$\begin{cases} \widehat{\delta}_{j+1} = \delta_j + (\omega_j - 1)\Delta t \\ \widehat{\omega}_{j+1} = \omega_j + \frac{P_m}{M}\Delta t - \frac{E'U}{MX_\Sigma}\sin\delta_j\Delta t - \frac{D}{M}(\omega_j - 1)\Delta t + \frac{\sigma_0}{M}\Delta B_j \\ \delta_{j+1} = \delta_j + \frac{1}{2}[(\omega_j - 1) + (\widehat{\omega}_j - 1)]\Delta t \\ \omega_{j+1} = \omega_j + \dfrac{P_m}{M}\Delta t - \dfrac{1}{2} \times \dfrac{E'U}{MX_\Sigma}(\sin\delta_j + \sin\widehat{\delta}_j)\Delta t \\ \qquad - \dfrac{1}{2} \times \dfrac{D}{M}[(\omega_j - 1) + (\widehat{\omega}_j - 1)]\Delta t + \dfrac{\sigma_0}{M}\Delta B_j \end{cases} \tag{3.4.2}$$

For a positive integer N against the system linear stochastic differential equation model (3.3.10), if:

$$\Delta t = (T - t_0)/N,\ \tau_j = j\Delta t,\ \Delta\omega_j = \Delta\omega(\tau_j),\ \Delta\delta_j = \Delta\delta(\tau_j) \text{ and } j = 0, 1, 2, \ldots, N$$

There are the following iteration formulas:

$$\begin{cases} \Delta\widehat{\delta}_{j+1} = \Delta\delta_j + \Delta\omega_j \Delta t \\ \Delta\widehat{\omega}_{j+1} = \Delta\omega_j - \frac{E'U}{MX_\Sigma}\cos\delta_0 \cdot \Delta\delta\Delta t - \frac{D}{M}\Delta\omega_j\Delta t + \frac{\sigma_0}{M}\Delta B_j \\ \Delta\delta_{j+1} = \Delta\delta_j + \frac{1}{2}(\Delta\omega_j + \Delta\widehat{\omega}_j)\Delta t \\ \Delta\omega_{j+1} = \Delta\omega_j - \dfrac{1}{2} \times \dfrac{E'U}{MX_\Sigma}\cos\delta_0 \cdot (\Delta\delta_j + \Delta\widehat{\delta}_j)\Delta t \\ \qquad - \dfrac{1}{2} \times \dfrac{D}{M}(\Delta\omega_j + \Delta\widehat{\omega}_j)\Delta t + \dfrac{\sigma_0}{M}\Delta B_j \\ \delta = \delta_0 + \Delta\delta_{j+1} \\ \omega = \omega_0 + \Delta\omega_{j+1} \end{cases} \quad (3.4.3)$$

3.4.2 Numerical Computation Stability

The stability of numerical methods is one of the key topics in stochastic differential analysis. In this section, the stability analysis of numerical methods is given for the numerical methods mentioned. Before proving the numerical stability, two inequalities, listed in the following lemmas, are introduced.

Lemma 3.1 Chebyshev inequality

Provided the expectation $E(\mathbf{X})$ *and variance* $D(\mathbf{X})$ *of the stochastic variable* $\mathbf{X}$ *are both present, for any* $\varepsilon > 0$, *the inequality*

$$P\{\|\mathbf{X} - E(\mathbf{X})\|_2 \geq \varepsilon\} \leq \frac{D(\mathbf{X})}{\varepsilon^2} \quad (3.4.4)$$

makes sense. According to Lemma 3.1, the smaller the variance of the stochastic variable X is, the greater the probability that X gets a value near the mean is.

Lemma 3.2 Gronwall inequality

Let integrable functions $\boldsymbol{\alpha}(t), \boldsymbol{\beta}(t) : [t_0, T] \to R$, *and L be a non-negative number, for any* $t \in [t_0, T]$, *if the following inequality is satisfied*

$$0 \leq \boldsymbol{\alpha}(t) \leq \boldsymbol{\beta}(t) + L\int_{t_0}^{T} \boldsymbol{\alpha}(s)\mathrm{d}s$$

Then

$$\boldsymbol{\alpha}(t) \leq \boldsymbol{\beta}(t) + L\int_{t_0}^{t} \mathrm{e}^{L(t-s)}\boldsymbol{\beta}(s)\mathrm{d}s \tag{3.4.5}$$

Theorem 3.3 *For stochastic differential Eq. (3.3.7), if the following conditions are met:*

(1) *Measurability:* $\mathbf{f}(\mathbf{X}(t), t)$ *and* $\mathbf{G}(\mathbf{X}(t), t)$, $(\mathbf{X}(t), t) \in$, $[t_0, T] \times R^n$ *and* L^{n+1} *are measurable;*
(2) *Lipschitz conditions: For any* $t \in [t_0, T], \mathbf{X}(t), \mathbf{Y}(t) \in R^n$, *there is* $K > 0$, *making*

$$\begin{aligned} \|\mathbf{f}(\mathbf{X}(t), t) - \mathbf{f}(\mathbf{Y}(t), t)\|_2 &\leq K\|\mathbf{X}(\mathbf{t}) - \mathbf{Y}(t)\|_2 \\ \|\mathbf{G}(\mathbf{X}(t), t) - \mathbf{G}(\mathbf{Y}(t), t)\|_2 &\leq K\|\mathbf{X}(\mathbf{t}) - \mathbf{Y}(t)\|_2; \end{aligned} \tag{3.4.6}$$

(3) *Linear growth conditions: For any* $t \in [t_0, T], \mathbf{X}(t) \in R^n$, *there is* $K > 0$, *making*

$$\begin{aligned} \|\mathbf{f}(\mathbf{X}(t), t)\|_2^2 &\leq K^2(1 + \|\mathbf{X}(t)\|_2^2) \\ \|\mathbf{G}(\mathbf{X}(t), t)\|_2^2 &\leq K^2(1 + \|\mathbf{X}(t)\|_2^2); \end{aligned} \tag{3.4.7}$$

(4) *Initial conditions:* $\mathbf{X}(t_0)$ *is* σ *—algebra measurable, and* $E(\|\mathbf{X}(t_0)\|_2^2) < \infty$.

In any finite time interval $[t_0, T]$, *the calculation method is numerically stable, that is, there is a* $\Delta_0 > 0$, *making* $\forall \in\ > 0,\ \Delta \in (0, \Delta_0)$, *and*

$$\lim_{\|\mathbf{X}_0^\Delta - \bar{\mathbf{X}}_0^\Delta\|_2 \to 0} \sup_{t_0 \leq t \leq T} P(\left\|\mathbf{X}_{n_t}^\Delta - \bar{\mathbf{X}}_{n_t}^\Delta\right\|_2 > \in) = 0 \tag{3.4.8}$$

where $\mathbf{X}_{n_t}^\Delta$ *represents the discrete value of the exact solution with a step length of* Δ, *and the initial value is* $\mathbf{X}^\Delta(t_0) = \mathbf{X}_0^\Delta$. $\bar{\mathbf{X}}_{n_t}^\Delta$ *represents the EM numerical solution of the same step length, and the initial value is* $\bar{\mathbf{X}}^\Delta(t_0) = \bar{\mathbf{X}}_0^\Delta$.

Proof According to the Chebyshev inequality from Lemma 3.1

$$P(\left\|\mathbf{X}_{n_t}^\Delta - \bar{\mathbf{X}}_{n_t}^\Delta\right\|_2 \geq \in) \leq \frac{1}{\in} E(\left\|\mathbf{X}_{n_t}^\Delta - \bar{\mathbf{X}}_{n_t}^\Delta\right\|_2^2)$$

Let $z(t) = \sup_{t_0 \leq t \leq T} E(\left\|\mathbf{X}_{n_t}^\Delta - \bar{\mathbf{X}}_{n_t}^\Delta\right\|_2^2)$, and according to condition (3) and the Gronwall inequality from Lemma 3.2, $z(t)$ satisfies

$$z(t) \leq \left\|\mathbf{X}_0^\Delta - \bar{\mathbf{X}}_0^\Delta\right\|_2^2 + K\int_0^t z(s)\mathrm{d}s \leq \left\|\mathbf{X}_0^\Delta - \bar{\mathbf{X}}_0^\Delta\right\|_2^2 + K\int_0^t \mathrm{e}^{K(t-s)}\left\|\mathbf{X}_0^\Delta - \bar{\mathbf{X}}_0^\Delta\right\|_2^2 \mathrm{d}s$$
$$= \left\|\mathbf{X}_0^\Delta - \bar{\mathbf{X}}_0^\Delta\right\|_2^2 \mathrm{e}^{Kt}$$

where $K > 0, t \in [t_0, T]$. When $\left\|\mathbf{X}_0^\Delta - \mathbf{X}_0^\Delta\right\|_2 \to 0$, it can be obtained that

$$\lim_{\left\|\mathbf{X}_0^\Delta - \bar{\mathbf{X}}_0^\Delta\right\|_2 \to 0} \sup_{t_0 \leq t \leq T} P(\left\|\mathbf{X}_{n_t}^\Delta - \bar{\mathbf{X}}_{n_t}^\Delta\right\|_2 > \in) \leq 0$$

Then it can be obtained by the nature of probability that

$$\lim_{\left\|\mathbf{X}_0^\Delta - \bar{\mathbf{X}}_0^\Delta\right\|_2 \to 0} \sup_{t_0 \leq t \leq T} P(\left\|\mathbf{X}_{n_t}^\Delta - \bar{\mathbf{X}}_{n_t}^\Delta\right\|_2 > \in) \geq 0$$

So

$$\lim_{\left\|\mathbf{X}_0^\Delta - \bar{\mathbf{X}}_0^\Delta\right\|_2 \to 0} \sup_{t_0 \leq t \leq T} P(\left\|\mathbf{X}_{n_t}^\Delta - \bar{\mathbf{X}}_{n_t}^\Delta\right\|_2 > \in) = 0$$

That is, the numerical method is numerically stable.

From Theorem 3.3, we can see that as long as the proper Δ_0 can be obtained, the EM method can be numerically stable. In other words, as long as the time step for numerical calculation is selected properly, the stability of such calculation can be guaranteed.

3.4.3 Model Comparison

In order to correctly describe the nature of the power system, ensure that the system model can represent the response characteristics of the actual system accurately, and ensure the validity of the conclusions, it is necessary to select the appropriate power system model for calculation. In this section, the system model that meets the subsequent calculation requirements is determined through the contrast of numerical calculations of the two system models under different stochastic disturbance intensities, based on the Heun numerical method iterative formula for nonlinear and linearized system models as shown in Eqs. (3.3.12) and (3.3.13). For the OMIB system with the system damping, the parameters used in simulation calculations are set as follows:

Step-up transformer $x_{T1} = 0.138$, double loop total reactance $x_l = 0.243$, step-down transformer $x_{T2} = 0.122$; generator transient reactance $x_d' = 0.295$,

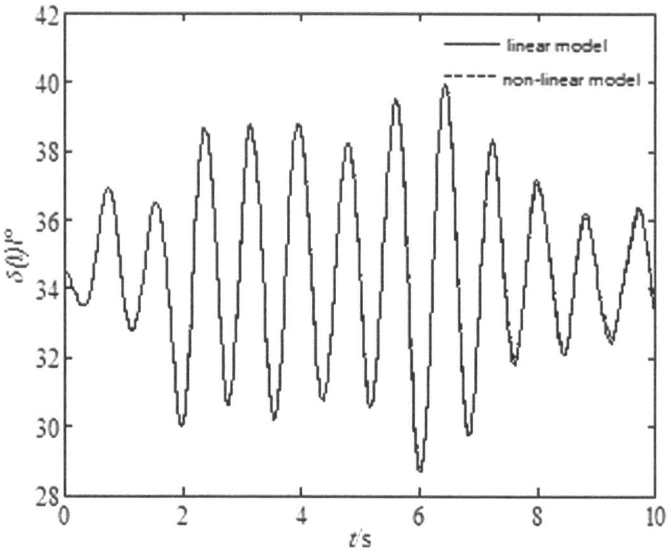

Fig. 3.13 Comparison of numerical results of power system model under a small stochastic disturbance

inertia time constant $M = 2569.8288$ (i.e. 8.18 s), damping coefficient $D = 2.0$; initial operating points $P_0 = 1.0$ and $Q_0 = 0.2$, electromotive force $E' = 1.41$, rotor angle $\delta_0 = 34.46°$. The above parameters are all per-unit values and their reference values are $S_B = 220$ MVA and $U_{B(220)} = 209$ kV.

When the stochastic disturbance intensity is $\sigma_0 = 0.1$, the numerical calculation result is shown in Fig. 3.13. It can be seen that the response curves of nonlinear and linearized stochastic differential equations basically coincide with each other under a small stochastic disturbance, and the degree of fitting between them is high. At this time, because the iterative solution formula of the linear system is simpler and less computational, and even for some specific systems, the analytical expression of the numerical solution can be written directly, which is beneficial to theoretical analysis. Therefore, a linearized stochastic differential equation model can be selected under the stochastic disturbance of low intensity.

When the stochastic disturbance intensity is $\sigma_0 = 0.66$, the numerical calculation result is shown in Fig. 3.14. It can be seen that under the stochastic large disturbance, the calculation results of the nonlinear model show that the system has lost stability at this time, but the calculation results of the linearization model is not the case, and the attenuation characteristics have even been found in the rotor angle oscillation of the system. This could be due to the neglect of the system's own nonlinear characteristics which result in a more obvious error in the calculation results. Therefore, under the strong stochastic disturbance, the linear stochastic differential equation model does not apply. As is needed to analyze the influence of stochastic disturbance intensity on the dynamic response characteristics of the

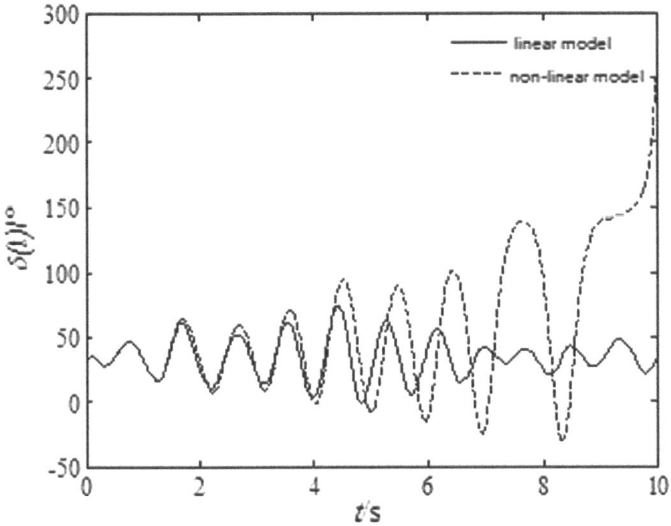

Fig. 3.14 Comparison of numerical results of power system model under a large stochastic disturbance

system for the subsequent study, it is more appropriate to select the nonlinear stochastic differential equation model.

3.4.4 Influence Factor Analysis

This section considers the nonlinear vector stochastic differential Eq. (3.3.7) and uses the iterative formula (3.3.12) of the Heun numerical method to calculate and analyze the effects of stochastic power disturbances on system stability. The effect of stochastic disturbances on the OMIB system is mainly determined by two parameters:

(1) Stochastic disturbance intensity σ_0: its size indicates the possible size of the stochastic disturbance. The stochastic disturbance received by the system can be regarded as a zero-mean Gaussian white noise process with smooth independent increments, which cannot directly represent the impact of stochastic disturbances intensity on the system. Therefore, the parameter σ_0 representing the stochastic disturbance intensity is introduced into the system model equation, and by adjusting such value, the maximum intensity of the stochastic power disturbance received by the power system can be considered.
(2) Disturbance step length dt: its size represents the time interval of the Gaussian-type stochastic power disturbance. Since it is assumed that two adjacent stochastic points before and after one disturbance step length are not

correlated, the disturbance step length is somewhat similar to the disturbance frequency and can be represented by the number of times of a system being impacted by stochastic power disturbances in a certain period of time.

(1) Impact of Stochastic Disturbance Intensity

The traditional power system stability refers to the ability of the power system to recover to the original steady-state operation mode or to achieve a new steady-state operation mode after disturbances, by means of the inherent capabilities of the system and the role of the control device. However, the stochastic power disturbance is essentially a stochastic process which has always existed in the system and continuously affected the dynamic response of the power system. Therefore, different from the common power system disturbances, the impact of the stochastic disturbance on the system needs a new method of stability discrimination.

In order to study the influence of this kind of stochastic disturbance, we consider that the dynamic response of the system should be tested several times at the same stochastic disturbance intensity σ_0, and the disturbance step length is fixed to be $\mathrm{d}t = 0.01$ s. When the stochastic disturbance intensity is $\sigma_0 = 0.8$, three stochastic calculation tests are performed respectively, and the system power angle response curve is shown in Fig. 3.15. In Fig. 3.15a, the system power angle gradually decreases from 7 to 10 s, and the response curve shows a damped oscillation phenomenon, indicating that the system gradually returns to stability within 7–10 s. In Fig. 3.15b, the system power angle gradually increases from 7 to 10 s, and the response curve shows a phenomenon of amplitude-increasing oscillation. The maximum value of power angle reaches about 80° and the system has approached the critical point. And if such an oscillation trend continues to run, instability will occur. In Fig. 3.15c, the system has an amplitude-increasing oscillation mode within 0–8 s. And at 8 s, the system is under periodical destabilization, the power angle increases sharply, and the power angle curve no longer swings back.

According to the results of Fig. 3.15, we can see that due to the stochasticity of Gaussian-type disturbances, the conclusions of system stability obtained under different stochastic disturbance tests are not unique, and may even be the opposite. Therefore, it is no longer applicable to judge whether the system is stable according to the system rotor angle response curve under a single stochastic disturbance, so we need to reconsider the criterion of system stability. In this section, the rotor angle mean curve under multiple stochastic disturbance tests is used as a criterion.

Figure 3.16 shows the system with 300 Gaussian stochastic disturbance tests with a stochastic disturbance intensity of $\sigma_0 = 0.8$, in which the system rotor angle response mean curve and two rotor angle response curve samples are plotted. It can be seen from the figure that: (1) There are obvious differences between the fluctuation laws of the stochastically selected two rotor angle response curves. This indicates that it is impossible to judge whether the system is stable based on a single rotor angle response trajectory because of the stochasticity of the system input; (2) The system rotor angle mean curve shows an overall trend of growth, and the mean value of the rotor angle increases obviously in the 8–10 s. This shows that in the 300 stochastic disturbance tests upon the system, most of the rotor angle

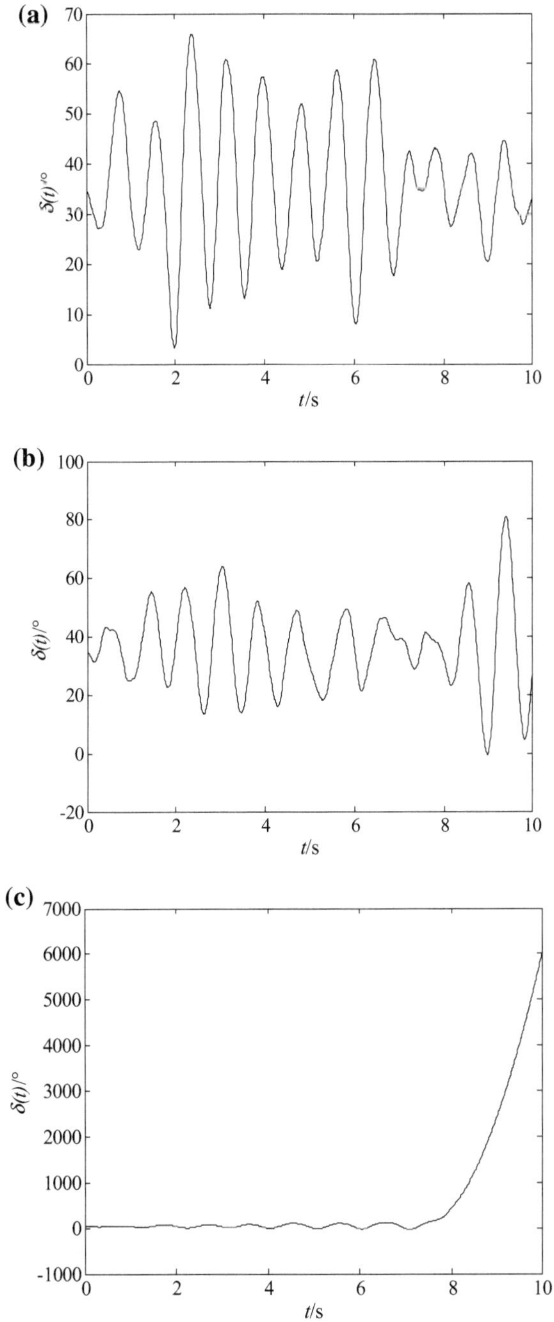

Fig. 3.15 System rotor angle response curves with stochastic disturbance intensity of 0.8

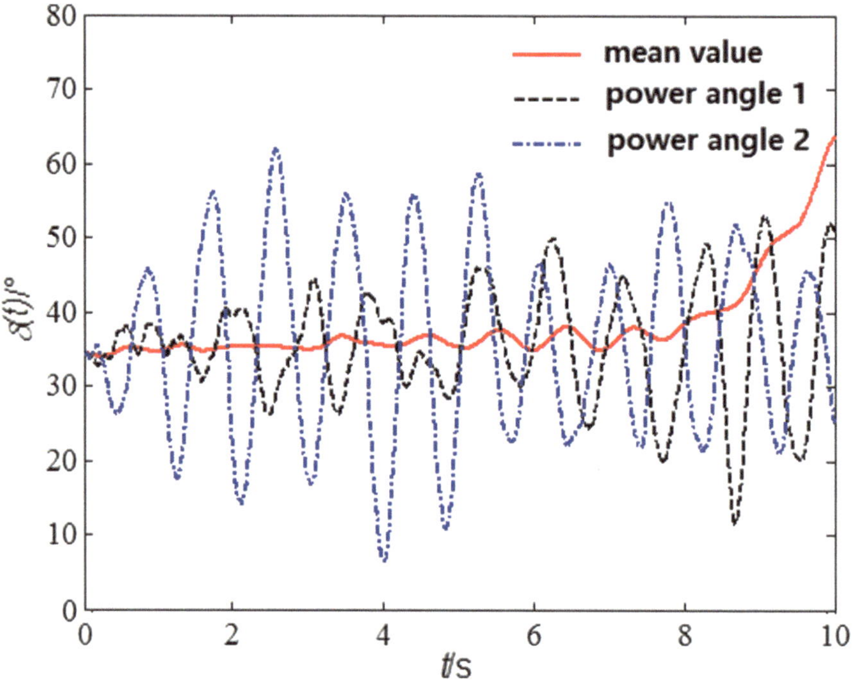

Fig. 3.16 System rotor angle response mean curve with stochastic disturbance intensity of 0.8

response curves show amplitude-increasing oscillation patterns, and the phenomenon of instability occurs within 8–10 s. It can be concluded that the OMIB system is unstable when the stochastic disturbance intensity is $\sigma_0 = 0.8$.

From the above analysis, it can be concluded that although the system has a trend of stabilization based on the simulation results of the system rotor angle response curve under a certain disturbance test, thereby the system can be judged to be stable. However, due to the stochasticity of the Gaussian white noise disturbance model, the system rotor angle response results also show a great stochasticity. When judging the stability of the system, it should be analyzed from the perspective of probability statistics, and based on the rotor angle mean curve, to determine whether the system is unstable under multiple stochastic disturbance tests. Subsequently, this section calculates the simulation results for 300 stochastic power disturbance tests under the condition that the value of the stochastic disturbance intensity σ_0 increases from 0.1 to 0.9. And by judging the stability of the rotor angle mean curve, the critical stochastic disturbance intensity value that leads to system instability is obtained.

Figure 3.17 shows the simulation results of the system response at a critical stochastic disturbance intensity of $\sigma_0 = 0.77$. It can be seen that the system mean rotor angle curve shows a significant increase trend at this time, which means that the system is unstable at this point. In summary, with the continuous increase of the

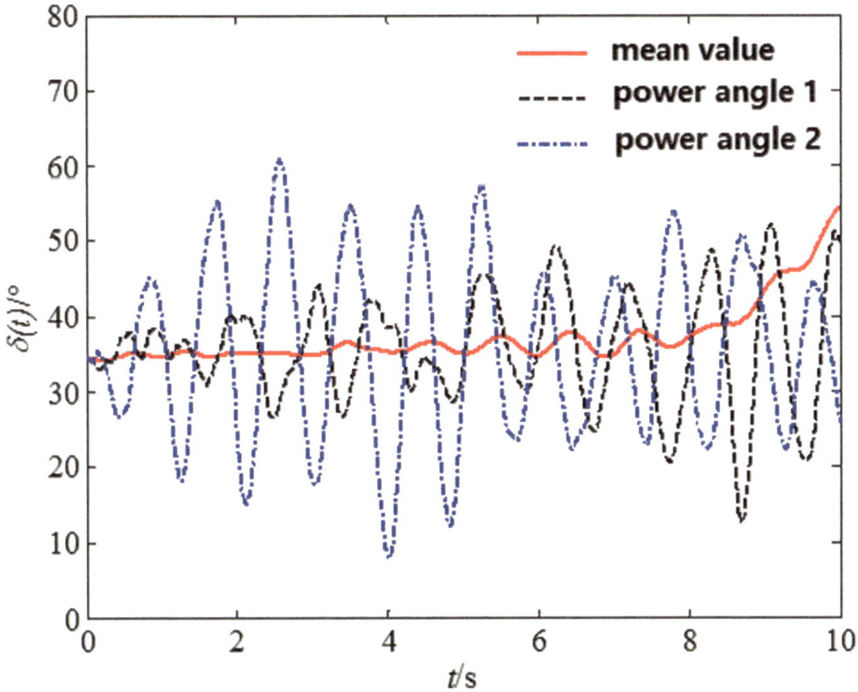

Fig. 3.17 System rotor angle mean curve under critical stochastic disturbance intensity

stochastic disturbance intensity σ_0, the range of oscillation amplitude of the system rotor angle curve is also increasing, and the system is destabilized at a certain critical intensity. For the example system of this chapter, it can be considered that 0.77 is the critical disturbance intensity that the system can withstand.

According to the OMIB model shown in Eq. (3.3.6), the stochastic disturbance is directly applied to the per-unit model equation in the form of an additive Gaussian white noise process $\sigma_0 W(t)$, and the relatively greater possible fluctuation amplitude of the standard Gaussian process is about −3 to 3, that is, the maximum stochastic disturbance amplitude that the system can withstand is about 2.31 times the power steady state value. In general, such high-intensity stochastic disturbance will not occur in an actual power system, but may occur in distributed microgrids or islanded power systems. In the power grid with current stochastic disturbance intensity, the stability problem will generally not occur in the system. However, the stochastic disturbance may cause the unknown oscillation of the system, which is worthy of further research.

(2) Impact of Stochastic Disturbance Step Length

In this section, the characteristics of the system rotor angle response curve for different disturbance step lengths of dt are analyzed, and the stochastic disturbance intensity $\sigma_0 = 0.1$ is selected for this purpose. Due to the stochasticity of

Gaussian-type disturbances, in order to ensure the comparability of the system response results, a fixed set of Gaussian-type stochastic disturbance data is used for calculation. At the same time, when the disturbance step length changes, it means that the time interval between adjacent stochastic disturbance points has changed, but the original Gaussian-type stochastic disturbance data points remain unchanged, so the total calculation time needs to change accordingly. Therefore, it is necessary to linearly interpolate the original data to increase the number of iteration points in order to ensure the smooth progress of the iterative process.

Figure 3.18 shows Gaussian-type stochastic disturbances with disturbance step lengths dt taking 0.01, 0.08, and 0.2 s, respectively. When the disturbance step length increases from 0.01 to 0.08 s and 0.2 s, the total simulation time increases proportionally in turn, and the time domain images of the three Gaussian-type stochastic disturbances stay the same, indicating that the ways the system is disturbed in the three tests are the same, but the frequencies of the disturbance have changed. After constructing a Gaussian-type stochastic disturbance model with different disturbance step lengths, perform simulations with the constructed stochastic disturbance as the system input, and analyze the properties of the system power-angle response curve. Figure 3.19 shows the time domain curves of the rotor angle response of the system under different disturbance step lengths.

It can be seen that due to the existence of stochastic disturbances and the constant change of the system rotor angle, no suitable conclusions on system responses can be drawn from the perspective of time domain analysis. Therefore, the frequency domain analysis on the rotor angle response is further performed. From the system response power spectra at different stochastic disturbance step lengths shown in Fig. 3.20, it can be seen that the energy of the signal is mainly concentrated at about 1.2 Hz and the signal energy at the remaining frequency

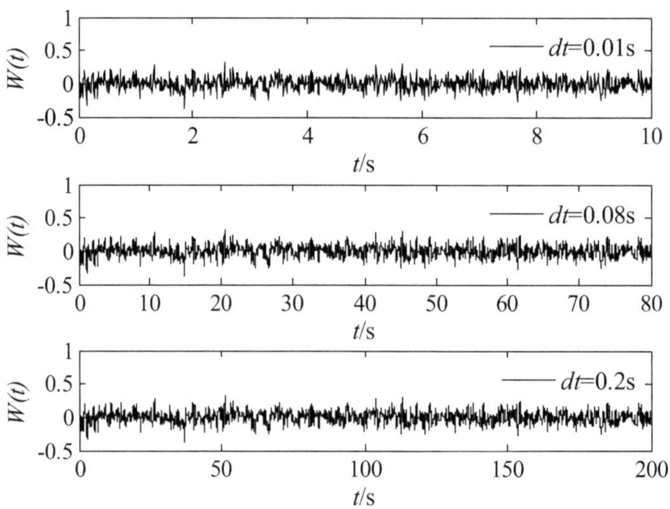

Fig. 3.18 Gaussian stochastic disturbances with different disturbance step lengths

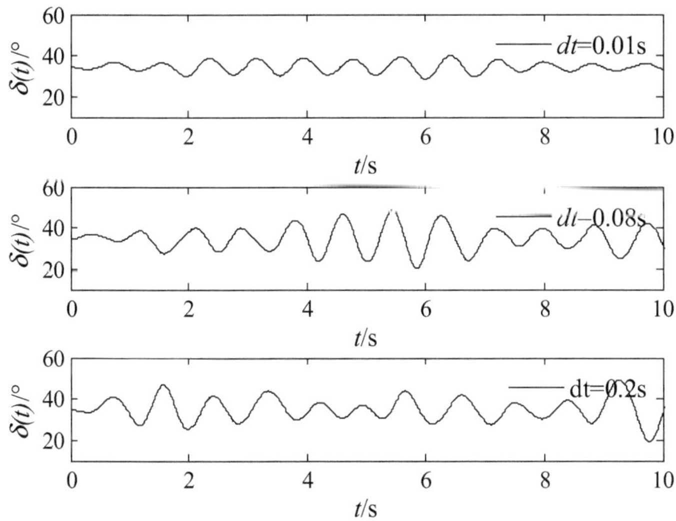

Fig. 3.19 Response curves of corresponding system rotor angle under stochastic disturbances with different disturbance step lengths

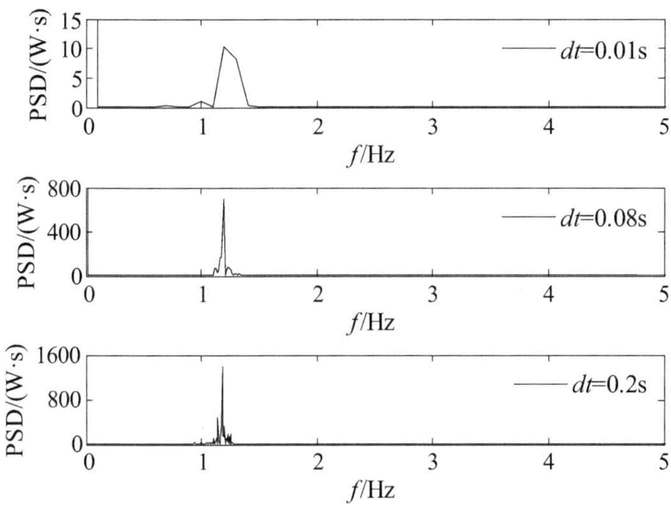

Fig. 3.20 System response time-domain curves corresponding power spectra

points is very small although the system response curves have no obvious regularity in the time domain diagram 3-19.

Table 3.1 lists the corresponding peak frequency values in the power spectra of the three different disturbance step lengths. By combining the power spectra of Fig. 3.20 and the peak frequencies shown in Table 3.1, it can be seen that the

Table 3.1 Power spectrum peak frequencies under different disturbance step lengths

Disturbance step length (s)	Peak frequency (Hz)
0.01	1.197
0.08	1.187
0.2	1.192

system output signal energy is concentrated at a frequency of about 1.19 Hz, and the signal energy at the remaining frequencies is very small.

For the single-machine infinite bus system, the system natural oscillation frequency ω_0 without damping has the following calculation formula [16]:

$$\omega_0 = \sqrt{\frac{E'U}{MX_\Sigma}\cos\delta_0} \tag{3.4.7}$$

By substituting the corresponding parameter values set in the system example in this chapter into formula (3.4.7), $\omega_0 \approx 0.0238$ (per-unit value) is calculated therefrom. With power frequency of 50 Hz, the natural oscillation frequency of the system can be calculated to be $f \approx 1.1905$ Hz, and this result is consistent with the peak frequency of the system response power spectrum shown in Table 3.1. It can be concluded that under Gaussian-type stochastic disturbances, i.e., when the system input is a stochastic signal, the system's rotor angle response curve is an oscillation curve, and its oscillation frequency is consistent with the natural oscillation frequency of the system. The Gaussian-type stochastic disturbances of different frequencies (step lengths) excite rotor angle oscillations of the same frequency (natural oscillation frequency) in the power system. This phenomenon exists in our current simulations, and the theoretical analysis for this phenomenon will be presented in Chap. 5.

References

1. Zhu, Weiqiu and Guoqiang Cai. 2017. *Introduction to Stochastic Dynamics*. Beijing: Science Press (in Chinese).
2. Zhu, Weiqiu. 2003. *Hamiltonian Theoretical Framework of Nonlinear Stochastic Dynamics and Control*. Beijing: Science Press. (in Chinese).
3. Ju, Ping. 2010. *The Theory and Method of Power System Modeling*. Beijing: Science Press. (in Chinese).
4. Kundur, P. 1994. *Power System Stability and Control*. New York, USA: McGraw-Hill Inc.
5. Xue, Yusheng. 1999. *Quantitative Theory for Motion Stability*. Nanjing: Jiangsu Science and Technology Press. (in Chinese).
6. Yu, Yixin and Chengshan Wang. 1999. *Theory and Method of Power System Stability*. Beijing: Science Press. (in Chinese).
7. Lu, Qiang, Yuanzhang Sun, and Shengwei Mei. 2001. *Nonlinear Control Systems and Power System Dynamics*. Boston: Kluwer Academic Publishers.
8. Research Group of Strategy of Energy Field of China. 2009. Energy technologies roadmap to 2050 of China. Beijing: Science Press (in Chinese).

9. Zhou, Xiaoxin, Zongxiang Lu, Yingmei Liu, et al. 2014. Development models and key technologies of future grid in China. *Proceedings of the CSEE* 34 (29): 4999–5008. (in Chinese).
10. Ju, Ping and Daqiang Ma. 1990. Probabilistic analysis of power system stability. *Automation of Electric Power Systems* 14 (3): 18–23. (in Chinese).
11. Xue, Yusheng, Yongjun Wu, Yunyun Xie, et al. 2016. Power system stability analysis for intercurrent natural disasters. *Automation of Electric Power Systems* 40 (4): 10–18. (in Chinese).
12. Li, G., H. Yue, M. Zhou, et al. 2014. Probabilistic assessment of oscillatory stability margin of power systems incorporating wind farms. *International Journal of Electrical Power & Energy Systems* 58: 47–56.
13. Yuan, B., M. Zhou, G. Li, et al. 2015. Stochastic small-signal stability of power systems with wind power generation. *IEEE Transactions on Power Systems* 30 (4): 1680–1689.
14. Bie, Zhaohong and Xifan Wang. 1997. The application of Monte Carlo method to reliability evaluation of power systems. *Automation of Electric Power Systems* 21 (6): 68–75. (in Chinese).
15. Zhu, W.Q. 2006. Nonlinear stochastic dynamics and control in hamiltonian formulation. *Applied Mechanics Reviews* 59 (4): 230–248.
16. Chen, Lincong and Weiqiu Zhu. 2010. Feedback maximization of reliability of a simple power system under random perturbations. *Journal of Dynamics and Control* 8 (1): 19–23. (in Chinese).
17. Zhu, Weiqiu and Zuguang Ying. 2013. Advances in research on nonlinear stochastic optimal control of quasi-hamiltoniansystems. *Advances in Mechanics* 43 (1): 39–55.
18. Qin, Chao and Yixin Yu. 2014. Security region based probabilistic small signal stability analysis for power systems with wind power integration. *Automation of Electric Power Systems* 38 (10): 43–48. (in Chinese).
19. Huang, Yang, Wei Hu, Yong Min, et al. 2014. Risk-constrained coordinative dispatching for large-scale wind-storage system. *Automation of Electric Power Systems*, 38(9): 41–47 (in Chinese).
20. Li, Yining, Hao Wu, Huanhai Xin, et al. 2015. Power system probabilistic load flow based on generalized polynomial chaso methods. *Automation of Electric Power Systems* 39 (7): 14–20. (in Chinese).
21. Li, Feng, Lizi Zhang, Jun Shu, et al. 2014. A flexible stochastic optimal scheduling approach considering intermittent power and energy storage. *Automation of Electric Power Systems* 38 (5): 1–7. (in Chinese).
22. Hou, Kai, Yuan Zeng, Hongjie Jia, et al. 2015. Risk assessment and optimization method for Markov chain-based scheduling operations. *Automation of Electric Power Systems* 39 (19): 142–148. (in Chinese).
23. Jiang, Le, Junyong Liu, Zhenbo Wei, et al. 2015. Running state and its risk evaluation of transmission line based on Markov Chain Model. *Automation of Electric Power Systems*, 39 (13): 51–57 (in Chinese).
24. Zhang, Xi, Chongqing Kang, Ning Zhang, et al. 2014. Analysis of mid/long term random characteristics of photovoltaic power generation. *Automation of Electric Power Systems*, 38 (6): 6–13 (in Chinese).
25. Zhu, Weiqiu and Guoqiang Cai. 2015. Introduction to stochastic dynamics. Beijing: Science Press (in Chinese).
26. Li, Hongyu, Ping Ju, Yiping Yu, et al. 2015. Bounded fluctuations regions and analytic method of intra-region probability in power system under stochastic excitations. *Proceedings of the CSEE* 35 (14): 3561–3568. (in Chinese).
27. Kundur, P., J. Paserba, V. Ajjarapu, et al. 2004. Definition and classification of power system stability—IEEE/CIGRE joint task force on stability terms and definitions. *IEEE Transactions on Power Systems* 19 (2): 1387–1401.

28. Wang, K. and M.L. Crow. 2013. The Fokker-Planck equation for power system stability probability density function evolution. *IEEE Transactions on Power Systems* 28 (3): 2994–3001.
29. Dong, Z.Y., J.H. Zhao, and D.J. Hill. 2012. Numerical simulation for stochastic transient stability assessment. *IEEE Transactions on Power Systems* 27 (4): 1741–1749.
30. Higham, D.J. 2001. An algorithmic introduction to numerical simulation of stochastic differential equations. *Society for Industrial and Applied Mathematics* 43 (3): 525–546.
31. Chen, Daxuan, Yiping Yu, Ping Ju, et al. 2014. Modeling of special power load based on time-varying current injection method. *Electric Power Automation Equipment* 34 (3): 120–124. (in Chinese).
32. Liu, Yongfei, Ping Ju, Yusheng Xue, et al. 2014. Calculation analysis on power system characteristics under random excitation. *Automation of Electric Power Systems* 38 (9): 137–142. (in Chinese).
33. Klein, M., G.J. Rogers, and P. Kundur. 1991. A fundamental study of inter-area oscillations in power systems. *IEEE Transactions on Power Systems* 6 (3): 914–921.
34. Ju, Ping, Yongfei Liu, Hongyin Wang, et al. 2014. General forced oscillations of power systems. *Electric Power Automation Equipment* 34 (5): 1–6. (in Chinese).
35. Sanchez-Gasca, J.J., V. Vittal, M.J. Gibbard, et al. 2005. Inclusion of higher order terms for small-signal (modal) analysis: committee report-task force on assessing the need to include higher order terms for small-signal (modal) analysis. *IEEE Transactions on Power Systems*, 20(4): 1886–1904.
36. Chen, Lei, Yong Min, and Yiping Chen. 2014. Study on internal resonance in power system low frequency oscillation. *Power System Technology* 38 (1): 160–167. (in Chinese).
37. Editorial Committee. 2014. China Electric Power Encyclopedia. Beijing: China Electric Power Press (in Chinese).
38. Zhang, J.Y., P. Ju, Y.P. Yu, et al. 2012. Responses and stability of power system under small Gauss type random excitation. *Science China Technological Sciences*, 55(7): 1873–1880.
39. Zhou, Ming, Bo Yuan, Xiaoping Zhang, et al. 2014. Stochastic small signal stability analysis of wind power integrated power systems based on stochastic differential equations. *Proceedings of the CSEE* 34 (10): 1575–1582. (in Chinese).
40. Liu, Yanli and Yixin Yu. 2013. Probabilistic steady-state and dynamic security assessment of power transmission system. *Science China Technological Science* 43 (5): 523–532.
41. Zhao, Yuan, Jie Wang, Yanjiao Xiong, et al. 2016. A Gaussian mixture model of multivariable in power system reliability assessment. *Automation of Electric Power Systems*, 40(1): 66–71 + 80 (in Chinese).
42. Li, Hongyu, Ping Ju, Xinqi Chen, et al. 2015. Stochastic averaging method for quasi Hamiltonian system of multi-machine power systems. *Science China Technological Sciences* 45 (7): 766–772. (in Chinese).

Chapter 4
Stochastic Stability of Power System

Abstract Stability is a basic requirement of power system. The response of power system to a stochastic disturbance is also a stochastic process. At this point, stability conditions of the stochastic process are to be given by probability or statistics, and usually defined through the convergence of the norm of the dynamic response. Meanwhile, the convergence of the stochastic process can be defined in different senses, so can the stochastic stability. This chapter introduces several common definition modes (i.e., the mean stability, mean square stability, and asymptotic stability) to demonstrate associated theorems.

4.1 Introduction

The power system, a complex non-linear system, always faces various kinds of stochastic factors [1, 2], such as short circuit faults, and power fluctuations, etc. As renewable energy power generation is getting grid-connected and electric automobiles are widely accessing to grid, stochastic factors will increase. References and researches [3–5] have shown that stochastic factors are adversely impacting the power system operation and control. At present, it is difficult to put all these factors together for description or study how they are impacting the power system. As for researches of how stochastic factors affect a power system, domestic and oversea scholars mainly focus on probabilistic stability and system simulating calculation [6–9]. A few references [10] utilized a stochastic differential equation to establish models, in order to discuss the effects of stochastic initial conditions, stochastic parameters, and so forth on system voltages, active power, reactive power, and stability, etc.

Regards to this issue, the author believes it is an important way of studying the power system stochastic stability issue by: selecting the power fluctuation, a stochastic factor caused by new energies (such as wind power integration) and wide accesses of electric automobiles; taking it as a stochastic disturbance to the system; modeling a non-linear stochastic power system and a linear one; theoretically analyzing impacts from the stochastic disturbance on the system; and analyzing

P. Ju, *Stochastic Dynamics of Power Systems*, Power Systems,
https://doi.org/10.1007/978-981-13-1816-0_4

impacts from power fluctuations on system rotor angle in the sense of simulating calculation.

4.2 Mean Stability of Power System under Small Stochastic Disturbance

Stochastic stability [11–17] can be analyzed in two ways. One is a direct way based on known system solution processes, primarily applicable to those stochastic systems capable of obtaining the solution processes. The other one is an indirect way making use of a Lyapunov function, mainly applied to those stochastic systems failing to obtain the solution processes. In general, the direct method is mainly suitable for moment stability, i.e., the p order moment stability. Solution processes of a stochastic model include stochastic terms. To eliminate troubles brought by the stochastic terms during analysis, we first take the mean value for the solution process and analyze the mean value accordingly. For a linear stochastic power system, a main target of the study is the mean value stability (one-order moment) and the mean square stability (two-order moment). For a non-linear stochastic power system, due to absent of solution processes, a Lyapunov function may be constructed by using a non-linear model for studying the asymptotic stability of the system.

To facilitate the discussion of the stochastic system stability, the author here gives a brief introduction of associated principal mathematical fundamentals.

4.2.1 Associated Mathematical Fundamentals

Definition 4.1 Let $\mathbf{X}(t)$ be a stochastic process, if the limit of $E\|\mathbf{X}(t)\|_2$ is less than a positive constant c, then the system is regarded as mean stable. Mean stability refers to that a mean value of the system response motivated by a stochastic input is bounded.

Definition 4.2 Let $\mathbf{X}(t)$ be a stochastic process, if the limit of $E\left\|\mathbf{X}(t)\mathbf{X}^T(t)\right\|_2$ is less than a positive constant C, then the system is regarded as mean square stable. Mean square stability refers to that a mean square deviation of the system response motivated by a stochastic input is bounded.

Definition 4.3 Given that a stochastic process is $\mathbf{X}(t,\omega;\mathbf{X}_0,t_0)$, for every $\varepsilon > 0$ there exists a $\delta(\varepsilon) > 0$, if $\|\mathbf{X}_0\|_2 < \delta(\varepsilon)$, it gives

$$\lim_{t\to\infty} P\left\{\|\mathbf{X}(t,\omega;\mathbf{X}_0,t_0)\|_2 > \varepsilon\right\} = 0$$

So it is said the stochastic process $\mathbf{X}(t, \omega; \mathbf{X}_0, t_0)$ is weakly stochastic asymptotic stable.

Theorem 4.1 *Given that a n-order matrix A has eigenvalues* $\lambda_1, \lambda_2, \ldots, \lambda_n$*, there exists a positive constant C such that* $\left\| \mathrm{e}^{\mathbf{A}} \right\|_2 \le C\mathrm{e}^{\lambda}$*, wherein* $\lambda = max\{Re(\lambda_1), \ldots, Re(\lambda_k)\}$.

Proof Given that a function $f(x)$ at λ_i is expanded into:

$$f(x) = f(\lambda_i) + f'(\lambda_i)(x - \lambda_i) + \frac{1}{2!}f''(\lambda_i)(x - \lambda_i)^2 + \cdots + \frac{1}{n!}f^{(n)}(\lambda_i)(x - \lambda_i)^n + \cdots$$

Supposing that a matrix A is similar to a Jordan matrix $\mathbf{J}$, that is:

$$\mathbf{A} \sim \mathbf{J} = \begin{bmatrix} \mathbf{J}_1 & & \\ & \ddots & \\ & & \mathbf{J}_k \end{bmatrix}$$

Where $\mathbf{J}_i$ is a Jordan matrix block for $i = 1, 2, \ldots, k, k \le n$

$$\mathbf{J}_i = \begin{bmatrix} \lambda_i & 1 & & \\ & \lambda_i & \ddots & \\ & & \ddots & 1 \\ & & & \lambda_i \end{bmatrix}_{r \times r} = \lambda_i \mathbf{I} + \mathbf{H}$$

Where

$$\mathbf{H} = \begin{bmatrix} 0 & 1 & & \\ & 0 & \ddots & \\ & & \ddots & 1 \\ & & & 0 \end{bmatrix}, \quad \mathbf{H}^2 = \begin{bmatrix} 0 & 0 & 1 & \\ & 0 & \ddots & 1 \\ & & \ddots & 0 \\ & & & 0 \end{bmatrix}, \ldots, \mathbf{H}^{r-1} = \mathbf{0}.$$

By substitution of the equation $\mathbf{J}_i = \lambda_i \mathbf{I} + \mathbf{H}$ into the function $f(x)$ at λ_i

$$\begin{aligned} f(\mathbf{J}_i) &= f(\lambda_i)\mathbf{I} + f'(\lambda_i)(\mathbf{J}_i - \lambda_i \mathbf{I}) + \cdots + \frac{1}{n!}f^{(n)}(\lambda_i)(\mathbf{J}_i - \lambda_i \mathbf{I})^n + \cdots \\ &= f(\lambda_i)\mathbf{I} + f'(\lambda_i)\mathbf{H} + \cdots + \frac{f^{(n)}(\lambda_i)}{n!}\mathbf{H}^n + \cdots \end{aligned}$$

Given that $f(x) = \mathrm{e}^x$, therefore

$$\mathrm{e}^{\mathbf{J}_i} = \mathrm{e}^{\lambda_i}[\mathbf{I} + \mathbf{H} + \frac{1}{2!}\mathbf{H}^2 + \cdots] = \mathrm{e}^{\lambda_i}\mathbf{B}_i$$

Where

$$\mathbf{B}_i = \begin{bmatrix} 1 & 1 & \frac{1}{2!} & & \cdots & \frac{1}{(r-1)!} \\ & \ddots & \ddots & \ddots & & \vdots \\ & & \ddots & \ddots & \ddots & \\ & & & \ddots & \ddots & \frac{1}{2!} \\ & 0 & & & \ddots & 1 \\ & & & & & 1 \end{bmatrix},$$

Matrix A is similar to a Jordan matrix $\mathbf{J}$, that is $\mathbf{A} = \mathbf{PJP}^{-1}$, so

$$\begin{aligned} \|\mathbf{e}^{\mathbf{A}}\|_2 &= \|\mathbf{P}\mathbf{e}^{\mathbf{J}}\mathbf{P}^{-1}\|_2 \leq \|\mathbf{P}\|_2 \|\mathbf{P}^{-1}\|_2 \|\mathbf{e}^{\mathbf{J}}\|_2 \\ &= \|\mathbf{P}\|_2 \|\mathbf{P}^{-1}\|_2 [\mathrm{e}^{\lambda_1}\|\mathbf{B}_1\|_2 + \cdots + \mathrm{e}^{\lambda_k}\|\mathbf{B}_k\|_2] \\ &\leq C\mathrm{e}^{\lambda} \end{aligned}$$

In conclusion,

$$\|\mathbf{e}^{\mathbf{A}}\|_2 \leq C\mathrm{e}^{\lambda}.$$

Where in $C = \|\mathbf{P}\|_2 \|\mathbf{P}^{-1}\|_2 [\|\mathbf{B}_1\|_2 + \cdots + \|\mathbf{B}_k\|_2]$.

Theorem 4.2 *Given that* Λ *is an arbitrary n-order matrix, thus* $R(\mathbf{\Lambda}\mathbf{\Lambda}^{\mathrm{T}}) = R(\mathbf{\Lambda}^{\mathrm{T}}\mathbf{\Lambda}) = R(\mathbf{\Lambda})$.$R(\cdot)$ *indicates to a rank.*

Proof It is only proven that $R(\mathbf{\Lambda}^{\mathrm{T}}\mathbf{\Lambda}) = R(\mathbf{\Lambda})$. Given that $\boldsymbol{x} = (x_1, x_2, \ldots, x_n)^{\mathrm{T}}$

A necessary and sufficient condition for $\mathbf{\Lambda}^{\mathrm{T}}\mathbf{\Lambda}\boldsymbol{x} = \mathbf{0}$ is $\mathbf{\Lambda}\boldsymbol{x} = \mathbf{0}$, it indicates that the two equations have a same solution space. Therefore,

$$n - R(\mathbf{\Lambda}^{\mathrm{T}}\mathbf{\Lambda}) = n - R(\mathbf{\Lambda}).$$

that is:

$$R(\mathbf{\Lambda}^{\mathrm{T}}\mathbf{\Lambda}) = R(\mathbf{\Lambda}).$$

Therefore, it will be enough to be prove that $\mathbf{\Lambda}^{\mathrm{T}}\mathbf{\Lambda}\boldsymbol{x} = \mathbf{0}$ is equivalent to $\mathbf{\Lambda}\boldsymbol{x} = \mathbf{0}$.

(1) If $\mathbf{\Lambda}\boldsymbol{x} = \mathbf{0}$, left multiplying $\mathbf{\Lambda}^{\mathrm{T}}$ will give:

$$\mathbf{\Lambda}^{\mathrm{T}}\mathbf{\Lambda}\boldsymbol{x} = \mathbf{\Lambda}^{\mathrm{T}}\mathbf{0} = \mathbf{0}$$

(2) If $\mathbf{\Lambda}^{\mathrm{T}}\mathbf{\Lambda}\boldsymbol{x} = \mathbf{0}$, left multiplying on both sides $\boldsymbol{x}^{\mathrm{T}}$

$$\begin{aligned}
\boldsymbol{x}^{\mathrm{T}}\boldsymbol{\Lambda}^{\mathrm{T}}\boldsymbol{\Lambda}\boldsymbol{x} &= \mathbf{0}\\
(\boldsymbol{\Lambda}\boldsymbol{x})^{\mathrm{T}}\boldsymbol{\Lambda}\boldsymbol{x} &= \mathbf{0}\\
\langle \boldsymbol{\Lambda}\boldsymbol{x}, \boldsymbol{\Lambda}\boldsymbol{x}\rangle &= 0\\
\Lambda\boldsymbol{x} &= \mathbf{0}
\end{aligned}$$

4.2.2 Mean Stability of Linear Power System

The stochastic linear system vector form of the single-machine infinite-bus system can be expressed as:

$$\mathrm{d}\mathbf{X}(t) = \mathbf{A}\mathbf{X}(t)\mathrm{d}t + \mathbf{Q}\mathrm{d}B(t) \tag{4.2.1}$$

where the state variable is:

$$\mathbf{X}(t) = \begin{pmatrix} \Delta\delta \\ \Delta\omega \end{pmatrix}$$

the initial condition is:

$$\mathbf{X}(t_0) = \begin{pmatrix} 0 \\ 0 \end{pmatrix}$$

the state matrix is:

$$\mathbf{A} = \begin{pmatrix} 0 & 1 \\ a & b \end{pmatrix}$$

where

$$a = -\frac{P_{\max}}{M}\cos(\delta_0 - \upsilon),\ b = -\frac{D}{M},\ \mathbf{Q} = \begin{pmatrix} 0 \\ -\frac{\sigma}{M} \end{pmatrix}.$$

The stochastic linear system vector form of the multi-machine system can be expressed as:

$$\mathrm{d}\mathbf{X}(t) = \mathbf{A}\mathbf{X}(t)\mathrm{d}t + \mathbf{P}_L\mathrm{d}\mathbf{B}(t) \tag{4.2.2}$$

where

$$\mathbf{X}=\begin{bmatrix}\Delta\delta_1\\ \vdots\\ \Delta\delta_n\\ \Delta\omega_1\\ \vdots\\ \Delta\omega_n\end{bmatrix},\ \mathbf{A}=\begin{bmatrix} & \mathbf{0}_{n\times n} & & & \mathbf{I}_{n\times n} & \\ \frac{-K_{11}}{M_1} & \frac{-K_{12}}{M_1} & \cdots & \frac{-K_{1n}}{M_1} & \frac{-D_1}{M_1} & \\ \vdots & \vdots & \vdots & \vdots & & \ddots \\ \frac{-K_{n1}}{M_n} & \frac{-K_{n2}}{M_n} & \cdots & \frac{-K_{nn}}{M_n} & & \frac{-D_n}{M_n}\end{bmatrix}$$

where $K_{ij}=\frac{\partial P_{ei}}{\partial \delta_j}$, $\mathbf{I}_{n\times n}$ is an n-dimensional unit matrix. $\mathbf{P}_L=[P_{L1},P_{L2},\ldots,P_{Ln}]^{\mathrm{T}}$, $\mathbf{B}=\begin{bmatrix}\mathbf{0}_{n\times n} & \mathbf{0}_{n\times n}\\ \mathbf{0}_{n\times n} & \mathbf{B}_L\end{bmatrix}$, $\mathbf{B}_L=diag\left[\frac{1}{M_1},\frac{1}{M_2},\ldots,\frac{1}{M_n}\right]$.

Combining the models (4.2.1) and (4.2.2), linear system models of a power system may be converted into a uniform way below:

$$\mathrm{d}\mathbf{X}(t)=\mathbf{A}\mathbf{X}(t)\mathrm{d}t+\mathbf{Q}\mathrm{d}\mathbf{B}(t),t\in[0,+\infty) \tag{4.2.3}$$

The initial value $\mathbf{X}(0)=\mathbf{X}_0$ is bounded and independent of Wiener stochastic process $\mathbf{B}(t)$. $\mathbf{X}(t),\mathbf{Q}$ is a n-dimensional vector, and $\mathbf{A}$ is an n-order matrix.

Here is a mean stability theorem of a power system linearized model.

Theorem 4.3 *Based on the assumption that the formula (4.2.3) is a linear constant coefficient stochastic differential model of the power system, $\mathbf{X}(t)$ is its corresponding solution process. If the power system keeps stable under a small disturbance, it will be mean stable under a Gaussian stochastic disturbance. It means that there exists a positive constant c such that*

$$\lim_{t\to\infty}E\|\mathbf{X}(t)\|_2<c. \tag{4.2.4}$$

Proof if the power system is stable under a small disturbance, it indicates that all eigenvalues of a coefficient matrix A have negative real part, i.e.,

$$\lambda=max\{Re(\lambda_1),\ldots,Re(\lambda_k)\}<0$$

Now it is required to prove the existence of a $c>0$, which satisfies (4.2.4).

According to the existence condition of the solution process, the solution process of the model (4.2.3) can be explicitly expressed as follows[18–21]:

$$\mathbf{X}(t)=\mathrm{e}^{\mathbf{A}t}\mathbf{X_0}+\int_0^t\mathrm{e}^{\mathbf{A}(t-s)}\mathbf{Q}\mathrm{d}B(s) \tag{4.2.5}$$

Given that the coefficient matrix $\mathbf{A}$ has an eigenvalue $\lambda_j=\alpha_j+i\beta_j$, $\alpha_j<0$, $j=1,2,\ldots,n$. According to a Cauchy-Schwarz inequality:

$$(E\|\mathbf{X}(t)\|_2)^2 \leq E\|\mathbf{X}(t)\|_2^2 = E[\mathbf{X}^{\mathrm{T}}(t)\mathbf{X}(t)] \tag{4.2.6}$$

Taking X(t) of (4.2.5) into the right side of (4.2.6), one can get:

$$\begin{aligned}
E[\mathbf{X}^{\mathrm{T}}(t)\mathbf{X}(t)] &= E\left\{\left[\mathrm{e}^{\mathbf{A}t}\mathbf{X_0} + \int\limits_0^t \mathrm{e}^{\mathbf{A}(t-s)}\mathbf{Q}\mathrm{d}B(s)\right]^{\mathrm{T}}\left[\mathrm{e}^{\mathbf{A}t}\mathbf{X_0} + \int\limits_0^t \mathrm{e}^{\mathbf{A}(t-s)}\mathbf{Q}\mathrm{d}B(s)\right]\right\} \\
&= E\left\{\left[\left(\mathrm{e}^{\mathbf{A}t}\mathbf{X_0}\right)^{\mathrm{T}} + \int\limits_0^t \mathbf{Q}^{\mathrm{T}}\mathrm{e}^{\mathbf{A}^{\mathrm{T}}(t-s)}\mathrm{d}B(s)\right]\left[\mathrm{e}^{\mathbf{A}t}\mathbf{X_0} + \int\limits_0^t \mathrm{e}^{\mathbf{A}(t-s)}\mathbf{Q}\mathrm{d}B(s)\right]\right\}
\end{aligned}$$

After organizing, it can be rewritten as below:

$$\begin{aligned}
E[\mathbf{X}^{\mathrm{T}}(t)\mathbf{X}(t)] = {} & E[(\mathrm{e}^{\mathbf{A}t}\mathbf{X}_0)^{\mathrm{T}}\mathrm{e}^{\mathbf{A}t}\mathbf{X}_0] \\
& + E[(\mathrm{e}^{\mathbf{A}t}\mathbf{X}_0)^{\mathrm{T}}\int\limits_0^t \mathrm{e}^{\mathbf{A}(t-s)}\mathbf{Q}\mathrm{d}B(s)] \\
& + E[\int\limits_0^t \mathbf{Q}^{\mathrm{T}}\mathrm{e}^{\mathbf{A}^{\mathrm{T}}(t-s)}\mathrm{d}B(s)\mathrm{e}^{\mathbf{A}t}\mathbf{X}_0] \\
& + E[\int\limits_0^t \mathbf{Q}^{T}\mathrm{e}^{\mathbf{A}^{T}(t-s)}\mathrm{d}B(s)\int\limits_0^t \mathrm{e}^{\mathbf{A}(t-s)}\mathbf{Q}\mathrm{d}B(s)]
\end{aligned} \tag{4.2.7}$$

As for the first expectation in the formula (4.2.7), by using the definition of vector's 2-norm, as well as the property of stochastic variable that its expectation equals to itself, one can obtain:

$$E((\mathrm{e}^{\mathbf{A}t}\mathbf{X}_0)^{\mathrm{T}}\mathrm{e}^{\mathbf{A}t}\mathbf{X}_0) = \|\mathrm{e}^{\mathbf{A}t}\mathbf{X}_0\|_2^2 \tag{4.2.8}$$

As far as the second expectation is concerned, $B(t)$ is a Wiener process, $(\mathrm{e}^{\mathbf{A}t}\mathbf{X}_0)^{\mathrm{T}}\int_0^t \mathrm{e}^{\mathbf{A}(t-s)}\mathbf{Q}\mathrm{d}B(s)$ would be a Wiener process as well. The Wiener process has an expectation of 0, thus

$$E[(\mathrm{e}^{\mathbf{A}t}\mathbf{X}_0)^{\mathrm{T}}\int\limits_0^t \mathrm{e}^{\mathbf{A}(t-s)}\mathbf{Q}\mathrm{d}B(s)] = 0 \tag{4.2.9}$$

As for the third term of the right side in Eq. (4.2.7), similar to the formula (4.2.9), that is:

$$E[\int_0^t \mathbf{Q}^{\mathrm{T}} \mathrm{e}^{\mathbf{A}^{\mathrm{T}}(t-s)} \mathrm{d}B(s) \mathrm{e}^{\mathbf{A}t} \mathbf{X}_0] = 0 \tag{4.2.10}$$

As for the last expectation, by use of a real-valued function's stochastic integration of the Wiener process, it directly gives:

$$\begin{aligned} & E[\int_0^t \mathbf{Q}^{\mathrm{T}} \mathrm{e}^{\mathbf{A}^{\mathrm{T}}(t-s)} \mathrm{d}B(s) \int_0^t \mathrm{e}^{\mathbf{A}(t-s)} \mathbf{Q} \mathrm{d}B(s)] \\ & = \int_0^t \mathbf{Q}^{\mathrm{T}} \mathrm{e}^{\mathbf{A}^{\mathrm{T}}(t-s)} \mathrm{e}^{\mathbf{A}(t-s)} \mathbf{Q} \mathrm{d}s \\ & = \int_0^t \|\mathrm{e}^{\mathbf{A}(t-s)} \mathbf{Q}\|_2^2 \mathrm{d}s \end{aligned} \tag{4.2.11}$$

Substituting Eqs. (4.2.8)–(4.2.11) into the Eq. (4.2.7), a new equality can be acquired as shown in (4.2.12):

$$E[\mathbf{X}^{\mathrm{T}}(t)\mathbf{X}(t)] = \|\mathrm{e}^{\mathbf{A}t} \mathbf{X}_0\|_2^2 + \int_0^t \|e^{\mathbf{A}(t-s)} \mathbf{Q}\|_2^2 \mathrm{d}s \tag{4.2.12}$$

According to the norm compatibility of matrix and the Theorem 4.1, the first term in the formula (4.2.12) satisfies:

$$\|\mathrm{e}^{\mathbf{A}t} \mathbf{X}_0\|_2^2 \leq \|\mathrm{e}^{\mathbf{A}t}\|_2^2 \|\mathbf{X}_0\|_2^2 \leq m_0 c_0^2 \mathrm{e}^{2\lambda t} \tag{4.2.13}$$

where $m_0 = \|\mathbf{X}_0\|_2^2$ is a positive constant, and $\lambda = \lambda(\mathbf{A})$ indicates a real part of the maximum eigenvalue of the matrix A. The second term of formula (4.2.12) satisfies:

$$\begin{aligned} & \int_0^t \|\mathrm{e}^{\mathbf{A}(t-s)} \mathbf{Q}\|_2^2 \mathrm{d}s \leq \int_0^t \|e^{\mathbf{A}(t-s)}\|_2^2 \|\mathbf{Q}\|_2^2 \mathrm{d}s \\ & = m_1 c_1^2 \int_0^t \mathrm{e}^{2\lambda(t-s)} \mathrm{d}s \\ & = m_1 c_1^2 \frac{1}{2\lambda} (\mathrm{e}^{2\lambda t} - 1) \end{aligned} \tag{4.2.14}$$

where $m_1 = \|\mathbf{Q}\|_2^2$ is a positive constant.

Considering all eigenvalues of A have a negative real part, that is $\lambda < 0$. Formulas (4.2.13) and (4.2.14) are substituted into (4.2.12), thus

$$E[\mathbf{X}^{\mathrm{T}}(t)\mathbf{X}(t)] \le m_0 c_0^2 \mathrm{e}^{2\lambda t} + m_1 c_1^2 \frac{1}{2\lambda}(\mathrm{e}^{2\lambda t} - 1) \tag{4.2.15}$$

Taking account into the Eq. (4.2.15) in the (4.2.6), resolving the limit at both sides at same time:

$$\begin{aligned} (\lim_{t\to\infty} E\|\mathbf{X}(t)\|_2)^2 &\le \lim_{t\to\infty} E\|\mathbf{X}(\mathbf{t})\|_2^2 \\ &= \lim_{t\to\infty}[m_0 c_0^2 \mathrm{e}^{2\lambda t} + m_1 c_1^2 \frac{\mathrm{e}^{2\lambda t} - 1}{2\lambda}] \\ &= -\frac{m_1 c_1^2}{2\lambda} \end{aligned} \tag{4.2.16}$$

where, due to $\lambda < 0$, $-\frac{m_1 c_1^2}{2\lambda}$ is a positive real number.

Therefore, if the value of c is greater than $\sqrt{-\frac{m_1 c_1^2}{2\lambda}}$, such as $c = \sqrt{-\frac{m_1 c_1^2}{2\lambda} + 1}$, so that

$$\lim_{t\to\infty} E\|\mathbf{X}(t)\|_2 < c$$

It can be seen that the system is mean stable from Definition 4.1.

According to the Theorem 4.3, it can be concluded in 3 aspects as follows:

(1) Any value greater than $\sqrt{-\frac{m_1 c_1^2}{2\lambda}}$ can be taken as a dynamic boundary of the stochastic system. For the example system, if the system is stable subject to a small disturbance, and the input with small stochastic excitation is bounded, then the dynamic responses of the system is the curve of rotor angle with fluctuations. At the same time the mean value of rotor angle curve is bounded.
(2) Dynamic response of the system consists of two parts in the Eq. (4.2.5). The first term is caused by the system initial condition X_0, and the second term by the system stochastic disturbance term $\mathbf{Q}W(t)$. However, in the Formula (4–16), the system initial condition is not associated with $\mathbf{X_0} \equiv \mathbf{0}$. Therefore, a mean value boundary size (or value) of the dynamic response is only associated with the stochastic disturbance term $\mathbf{Q}W(t)$.
(3) The mean stability is visual and brief in describing the stability of a stochastic system. However, it has a certain defect. Theoretically, the boundary c should take a small value as possible. While practically seen from the derivation, c cannot take values precisely, because it depends on the amplification or reduction of the inequality. Sometimes it takes a value too large which is no longer suitable for stability description.

4.3 Mean Square Stability of Power System under Small Stochastic Disturbance

Mean stability describes that a dynamic response of the system is of a bounded expectation when it is subject to a stochastic disturbance. However, usually, this does not fully reflect the system stability. For example, when two different stochastic models are used to describe a same system, if bounded values of the mean stability differ little which can hardly tell differences between the two models. This is similar to expectations in probability statistics. What comes next is mean square stability [22–27] based on the variance principle.

Theorem 4.4 *Given that the formula (4.2.3) is a linear constant coefficient stochastic differential model of a power system,* $\mathbf{X}(t)$ *is its corresponding solution process. If the power system keeps stable under a small disturbance, it will be mean square stable under a Gaussian stochastic disturbance. That is, there is a positive number C so that*

$$\lim_{t\to\infty} E\left\|\mathbf{X}(t)\mathbf{X}^{\mathrm{T}}(t)\right\|_2 < C.$$

Proof As $\mathbf{X}(t)$ is a real matrix, and thus $\mathbf{X}(t)\mathbf{X}^{\mathrm{T}}(t)$ is a real matrix as well, so

$$\left[\mathbf{X}(t)\mathbf{X}^{\mathrm{T}}(t)\right]^{\mathrm{H}} = \left[\mathbf{X}(t)\mathbf{X}^{\mathrm{T}}(t)\right]^{\mathrm{T}} = \mathbf{X}(t)\mathbf{X}^{\mathrm{T}}(t)$$

That is, $\mathbf{X}(t)\mathbf{X}^{\mathrm{T}}(t)$ is a normal matrix. Based on the definition and property of 2-norm, it results in

$$\left\|\mathbf{X}(t)\mathbf{X}^{\mathrm{T}}(t)\right\|_2 = \rho\left(\mathbf{X}(t)\mathbf{X}^{\mathrm{T}}(t)\right)$$

According to Theorem 4.2

$$R\left(\mathbf{X}(t)\mathbf{X}^{\mathrm{T}}(t)\right) = R(\mathbf{X}(t)) = 1$$

That is, the n-order matrix $\mathbf{X}(t)\mathbf{X}^{\mathrm{T}}(t)$ has a rank of 1, and all its eigenvalues, except its trace $tr\left(\mathbf{X}(t)\mathbf{X}^{\mathrm{T}}(t)\right)$, are zero. As a result

$$\rho\left(\mathbf{X}(t)\mathbf{X}^{\mathrm{T}}(t)\right) = tr\left(\mathbf{X}(t)\mathbf{X}^{\mathrm{T}}(t)\right) = \mathbf{X}(t)\mathbf{X}^{\mathrm{T}}(t)$$

Therefore

$$\lim_{t\to\infty} E\left\|\mathbf{X}(t)\mathbf{X}^{\mathrm{T}}(t)\right\|_2 = \lim_{t\to\infty} E\left\|\mathbf{X}(t)\mathbf{X}^{\mathrm{T}}(t)\right\|$$

The next proof shares a same idea with the proof of Theorem 4.3, which takes $C = c^2$, so:

$$\lim_{t \to \infty} E\left\| \mathbf{X}(t)\mathbf{X}^{\mathrm{T}}(t) \right\|_2 < C$$

It can be concluded that the system is mean square stable from Definition 4.2. It can be found through analyzing the mean square stability of the power system subject to a Gaussian disturbance:

(1) Similar to the mean stability, if the system is stable under a small disturbance, the mean square boundary size (or value) of the dynamic response is only associated with a stochastic disturbance term $\mathbf{Q}W(t)$.
(2) To some extent, C can reflect how large a stochastic power system angle deviates from a mean value during fluctuation. As a result, C should be as small as possible.
(3) Two kinds of stability are put into comparison. Under a small stochastic disturbance, the mean stability mainly shows that if a dynamic mean is in the neighbourhood of a steady-state value. The smaller the neighbourhood radius c is, the closer the dynamic mean value is to the steady-state value. However, the mean square stability describes an amount by which a dynamic response deviates from the dynamic mean value. It will not be directly associated with the steady-state value. That is, to illustrate the effect of the mean square stability, the connection with the steady-state value through mean stabilization has to be established. Therefore, when analyzing moment stability of the system, the mean stability and mean square stability shall both be covered. Meanwhile, the mean stability acts as a precondition or foundation of the mean square stability.

4.4 Asymptotic Stability of Power System Under Large Stochastic Disturbance

Through the moment stability analyzed on the basis of the linearized power system model, it is concluded that: if the power system keeps stable under a small disturbance, it will be mean stable and mean square stable under a Gaussian stochastic disturbance. This analysis brings benefits by: subtly combining the small disturbance stability analysis in power system together with the moment stability analysis in the stochastic system. In contrast, it is disadvantageous in following aspects: (1) The linearized model can only replace a non-linearized model within the small neighborhood of the initial point. This method is not suitable in the case of strong disturbances. (2) Stability analysis relies too much on the system solution process. Analysis would not be carried out if no solution process is provided. (3) Although the solution process of the stochastic system is explicit, it includes integration of the Wiener process.

Accordingly, it is required to consider a non-linear model based on the power system so as to perform the stochastic stability analysis. However, stochastic stability analysis of a non-linear model is very difficult due to lack of a common approach. Next, taking a single-machine infinite bus (SMIB) system for instance, the author is going to theoretically analyze asymptotic stability problems happening to a non-linear stochastic model subject to a large stochastic disturbance. A simulation example is provided as well.

To discuss whether an SMIB non-linear system subject to a stochastic disturbance has weakly stochastic asymptotic stability, the author cited many related definitions, lemmas, and theorems in this part. These cited portions are complicated. To facilitate reading, they are not listed here one by one. Only citation sources are provided.

Content referring to the weakly stochastic asymptotic stability is partly cited from references [22]. Original authors have provided part of the proof. Content referring to the structure of a non-linear single-machine infinite bus system is from reference [6]. The sections below will prove a non-linear stochastic model, the Lyapunov function, and the weakly stochastic asymptotic stability of an SMIB system.

4.4.1 Theorem of Weakly Stochastic Asymptotic Stability

Suppose that a research system is:

$$\begin{cases} \frac{\mathrm{dx}}{\mathrm{d}t} = \mathbf{F}(\mathbf{X}, t) + \boldsymbol{\sigma}(\mathbf{X}, t)\xi(t, \omega) \\ \mathbf{F}(0, t) \equiv 0, \boldsymbol{\sigma}(0, t) \equiv 0 \end{cases} \tag{4.4.1}$$

The truncated system for the system (4.4.1) is:

$$\frac{\mathrm{d}\mathbf{X}}{\mathrm{d}t} = \mathbf{F}(\mathbf{X}, t) \tag{4.4.2}$$

where $\mathbf{X} = (X_1, X_2, \ldots, X_n)^{\mathrm{T}}$ is a system state variable.

Theorem 4.5 *For the truncated system (4.4.2), suppose that there exists a Lyapunov function* $\mathbf{V}(\mathbf{X}, t) \in \mathbf{C}_0$ *with respect to t, and satisfying the following conditions:*

(1) *The Lyapunov function V(X,t) is consistent and positive respect to the variable t which means for all t > 0, r > 0, the Lyapunov function satisfies:*

$$\inf_{t > 0, \|\mathbf{X}\|_2 > r} \mathbf{V}(\mathbf{X}, t) = \mathbf{V}_r > 0 \tag{4.4.3}$$

(2) *The Lipschitz condition, i.e.:*

$$|\mathbf{V}(\mathbf{X}_2, t) - \mathbf{V}(\mathbf{X}_1, t)| \leq B|\mathbf{X}_2 - \mathbf{X}_1|, \mathrm{B} > 0;$$

(3) *The initial condition is* $\mathbf{V}(0,0) \equiv 0$;

(4) $\frac{\mathrm{d}^{(0)}\mathbf{V}}{\mathrm{d}t} \leq \quad C_1\mathbf{V}$, *$C_1$ is a constant larger than 0;*

(5) $||\boldsymbol{\sigma}(\mathbf{X}, t)|| \leq C_2\mathbf{V}$, *the constant* $C_2 > 0$;

(6) *The stochastic process* $\|\xi(t,\omega)\|_2$ *satisfies the law of large numbers, and*

$$\sup_{t>0} E\|\xi(t,\omega)\|_2 < \frac{C_1}{BC_2} \tag{4.4.4}$$

Then the system (4.4.1) is weakly and asymptotically stochastic stable, which meets Definition 4.3.

4.4.2 The Lyapunov Function of SMIB System

It is assumed that the swing equations of the generator rotor is:

$$M\frac{\mathrm{d}^2\delta}{\mathrm{d}t^2} + D\frac{\mathrm{d}\delta}{\mathrm{d}t} = P_m - P_e \tag{4.4.5}$$

Let be $x_1 = \delta - \delta^*$, $x_2 = \dot{x}_1$, $\delta^* = \arcsin(\frac{P_m}{P_e})$, the mathematical model of the system can be written as:

$$\begin{cases} \dot{x}_1 = x_2 \\ \dot{x}_2 = -\frac{D}{M}x_2 - \frac{1}{M}f(x_1) \end{cases}$$

where $f(x_1) = P_e\sin(x_1 + \delta^*) - P_e\sin\delta^*$.

Given that

$$\mathbf{A} = \begin{pmatrix} 0 & 1 \\ 0 & -\frac{1}{M} \end{pmatrix}, \quad \mathbf{B} = \begin{pmatrix} 0 \\ \frac{1}{M} \end{pmatrix}, \quad \mathbf{C} = \begin{pmatrix} 1 \\ 0 \end{pmatrix},$$

it is noted that $\mathbf{C}^{\mathrm{T}}\mathbf{B} = \mathbf{0}$, thus it takes $\boldsymbol{w}_0 = 0$, $\boldsymbol{U}_0 = 0$, and as a result

$$\mathbf{P} = \begin{pmatrix} Dn & MD \\ Mn & Mq \end{pmatrix}, \quad \mathbf{L} = \begin{pmatrix} 0 \\ \sqrt{2(Dq - nM)} \end{pmatrix}$$

and

$$\begin{pmatrix} 0 & 0 \\ 0 & L_2^2 \end{pmatrix} - 2\begin{pmatrix} 0 & 0 \\ 0 & m \end{pmatrix} = \begin{pmatrix} H_1^2 & H_1H_2 \\ H_1H_2 & H_2^2 \end{pmatrix}$$

If let H_1 and H_2 be zero then

$m = L_2^2/2$. It is easy to verify

$$\mathbf{P} - \frac{mn}{q}\mathbf{CC}^{\mathrm{T}} \geq \mathbf{0}$$

So far, one can get the improved Lyapunov function of the OMIB system as:

$$\mathbf{V}(\boldsymbol{x}) = \frac{1}{2}\boldsymbol{x}^T\boldsymbol{P}\boldsymbol{x} + q\int_0^{\sigma} f(\boldsymbol{\sigma})\mathrm{d}\boldsymbol{\sigma} + 2\sqrt{n(Dq - mn)}\int_0^{\sigma} \sqrt{\boldsymbol{\sigma}f(\boldsymbol{\sigma})}^T \mathrm{d}\boldsymbol{\sigma} \tag{4.4.6}$$

For the SMIB system non-linear model, if the circuit resistance is further neglected, the corresponding truncated system shall be:

$$\begin{cases} \frac{\mathrm{d}\delta}{\mathrm{d}t} = \omega - 1 \\ \frac{\mathrm{d}\omega}{\mathrm{d}t} = \frac{P_m}{M} - \frac{E'U}{MX_\Sigma}\sin\delta - \frac{D}{M}(\omega - 1) \end{cases} \tag{4.4.7}$$

Corresponding to the Lyapunov function, the function is $f(t) = \frac{E'U}{X_\Sigma}\sin t - P_m$, the state variable is $\boldsymbol{x} = (\delta, \omega)^{\mathrm{T}}$, P is a square matrix, and q, D, m, n are all constants.

4.4.3 *Verification of Weakly Stochastic Asymptotic Stability*

The following verification is to be carried on to show that the Eq. (4.4.6) satisfy all six conditions that in the Theorem 4.5.

Conditions (1) and (4) have been proved in references [22]. It is easy to verify that $\mathbf{V}(\mathbf{X}, t)$ meets the vector form of Lipschitz condition. Similarly, so it is also easy to do that for Condition (2).

Condition (3) is obviously satisfied when the state variable X and time t are both 0.

In the system (4.4.7), $\sigma(X, t) = \frac{\sigma}{M}$ is a bounded constant, and it is easy to find a constant C_2 to satisfy Condition (5), such as when C_2 takes a greater value than $\left|\frac{\sigma_0}{M}\right|$ does, σ_0 indicates the largest disturbance strength.

With regard to (6), firstly, what the system (4.4.7) requires is that its stochastic processes $\xi(t, \omega)$ meet Condition (6), such as the Gaussian white noise. It is known from the Schwarz in equation that $E|\xi(t,\omega)|$ is bounded, the following inequality can be satisfied by adjusting the values of B, C_1, C_2

$$\sup_{t>0} E|\xi(t,\omega)| < \frac{C_1}{BC_2}$$

Secondly, what need to do is to prove that $\xi(t,\omega) - W(t)$ satisfies the law of large numbers. It would be sufficient, according to the definition of the law of large numbers, to prove:

$$\lim_{t\to\infty} P\{|\frac{1}{t}\int_{t_0}^{t+t_0} W(s)\mathrm{d}s - \frac{1}{t}\int_{t_0}^{t+t_0} EW(s)\mathrm{d}s| > \delta\} = 0$$

Considering the stochastic process $\eta(t) = \frac{1}{t}\int_{t_0}^{t+t_0} W(s)\mathrm{d}s$, as $\eta(t)$ is also a Gaussian stochastic process and its variance is bounded, it is denoted as $D[\eta(t)] = \sigma^2$, taking into consideration $EW(s) = 0$, the result can be achieved according to the Chebyshev inequality:

$$P\{|\frac{1}{t}\int_{t_0}^{t+t_0} W(s)\mathrm{d}s| > \delta\} = P\{|\eta(t)| > \delta t\} \le \frac{\sigma^2}{(\delta t)^2}$$

Thus

$$\lim_{t\to\infty} P\{|\eta(t)| > \delta t\} \le \lim_{t\to\infty} \frac{\sigma^2}{(\delta t)^2} = 0$$

Then, it can be seen from property of the probability that:

$$\lim_{t\to\infty} P\{|\eta(t)| > \delta t\} \ge 0$$

Therefore

$$\lim_{t\to\infty} P\{|\eta(t)| > \delta t\} = 0$$

Thus the condition (6) is met.

Conclusions can be drawn from this section: (1) Conditions (1) to Condition (4) are specific for the truncated system that is an original deterministic system. If these four conditions are satisfied, then original deterministic system is stable. While Condition (5) is specific for the disturbance strength, and Condition (6) is specific for the stochastic process. That is, if the original deterministic system is stable and the stochastic disturbance satisfies the conditions mentioned above, the stochastic system is weakly stochastic asymptotic stable. (2) The above conditions are sufficient but not necessary ones. In other words, if these conditions cannot be satisfied, the weakly stochastic asymptotic stability might not be guaranteed.

Conclusion for this Chapter: (1) No more stochastic stability problem will occur in the power system subject to a small stochastic disturbance, which means the original system will remain its stability or its instability. (2) Stochastic stability problems might happen to the power system subject to a large stochastic disturbance. This may occur in microgrids or small isolated grids. Although the stability problems are unlikely to occur in a large power grid, it is a question to study in the next chapter that whether a stochastic disturbance would cause a new stochastic oscillation problem, or a new resonance problem.

References

1. Kang, Chongqing, Qing Xia, and Wei Xu. 2011. Power System Uncertainty Analysis. Beijing: Science Press (in Chinese).
2. Ju, Ping, Chuan Qin, Hua Huang, et al. 2012. Research trends of power system modeling geared to smart grid. *Automation of Electric Power Systems* 36 (11): 1–6. (in Chinese).
3. Yu, Yixin, and Chengshan Wang. 1999. *Power System Stability Theory and Approach*. Beijing: Science Press. (in Chinese).
4. Kundur, P. 1994. *Power System Stability and Control*. NewYork: McGraw-Hill.
5. Ni, Yixin, Shousun Chen, and Baolin Zhang. 2002. *Dynamic Power System Theory and Analysis*. Beijing: Tsinghua University Press. (in Chinese).
6. Cao, Yijia, Shijie Cheng, and Deshu Chen. 1993. An improved Lure Lyapunov function and its applications in power system stability analysis. *Proceedings of the CSEE* 13 (6): 46–53. (in Chinese).
7. Lu, Qiang, Shengwei Mei, and Yuanzhang Sun. 2008. *Power System Non-linear Control*. Beijing: Tsinghua University Press. (in Chinese).
8. Ju, Ping. 2010. *Power System Modeling Theory and Method*. Beijing: Science Press. (in Chinese).
9. Odun-Ayo, T., and M.L. Crow. 2012. Structure-preserved power system transient stability using stochastic energy functions. *Power Systems, IEEE Transactions on* 27 (3): 1450–1458.
10. Zhang, Jianyong, Ping Ju, and Yiping Yu, et al. 2012. Responses and stability of power system under small Gauss type random excitation. *Sci China Tech Sci* 55 (7): 1873–1880.
11. Itô, K., and H.P. McKean. 1996. *Diffusion Processes and Their Sample Paths*. Berlin: Springer Verlag.
12. Itô, K. 1951. On stochastic differential equations. *Memoirs American Mathematical Society* 4: 1–51.
13. Stratonovich, R.L. 1966. A new representation for stochastic integrals and equations. *SIAM Journal on Control* 4 (2): 362–371.
14. Stratonovich, R.L. 1967. *Topics in the theory of random noise: general theory of random processes, nonlinear transformations of signals and noise*. Amsterdam: Gordon and Breach.
15. Arnold, L. 1998. *Random Dynamical Systems*. New York: Springer Verlag.
16. Zhang, Binggen, and Yuzhi Zhao. 1980. *Stochastic Differential Equations in Sciences and Engineerings*. Beijing: Ocean Press. (in Chinese).
17. Arnold, L. 1973. *Stochastic Differential Equations: Theory and Applications*. NewYork: Wiley.
18. Zhao, Ziren, and Xiuyan Peng. 2007. *Stochastic Process Fundamentals and Applications*. Harbin: Harbin Engineering University Press. (in Chinese).
19. Øksendal, B. 2003. *Stochastic differential equations*. Berlin Heidelberg: Springer.

20. Øksendal B. 2003. *Stochastic Differential Equations: An Introduction with Applications (Sixth ed)*. Berlin: Springer.
21. Friedman, A. 1975. *Stochastic Differential Equations and Applications*. New York: Academic Press.
22. Khasminskii, R. 2011. *Stochastic Stability of Differential Equations: Stochastic Modelling and Applied Probability*. New York: Springer.
23. Mao, X. 1997. *Stochastic Differential Equations and their Applications*. Chichester: Howood.
24. Kushner, H.J. 1967. *Stochastic Stability and Control*. London: Academic Press.
25. Kushner, H. 1972. Stochastic Stability. *Stability of Stochastic Dynamical systems*, 97–124.
26. Foss, S., and T. Konstantopoulos. 2004. An overview of some stochastic stability methods. *Journal of the Operations Research* 47 (4): 275–303.
27. Xuanda, Hu. 1986. *Stochastic Differential Equation Stability Theory*. Nanjing: Nanjing University Press. (in Chinese).

Chapter 5
Stochastic Dynamic Oscillation of Power System

Abstract Oscillation phenomena occur regularly in power grids worldwide. Recently some oscillation phenomena of unknown mechanism have frequently taken place, behaving highly stochastically. Meanwhile, at present, fast-growing renewable power generations and electrical vehicles have become grid-tied which would probably bring significant stochastic disturbances. Therefore, it is worth discussing and researching whether the stochastic disturbances can induce new oscillation phenomena in the power grids. This chapter describes at first a general forced oscillation phenomenon possibly caused by small stochastic disturbance, analyzes the mechanism thereof, and verifies the same using simulation examples and data on real power grids. Next, this chapter provides a general explanation of the internal resonance phenomenon presumably stimulated by large stochastic disturbance, clarifies the mechanism thereof and preliminarily verifies the same using simulation examples.

5.1 Introduction

As a common problem in large-scale interconnected power systems, the oscillation phenomena appear in worldwide power grids. It is mainly featured by oscillations of constant amplitude or amplification. Furthermore, there is an international development tendency in promoting the renewable energy industry and increasing the penetration rate of electrical vehicles. The obvious stochastic property, however, inevitably becomes a hazard affecting the security and stability of power grids. According to existing investigation data, recently some oscillation phenomena of unknown mechanism have frequently taken place, behaving highly stochastically. Currently, the grid tie of fast-growing renewable power generations and electrical vehicles probably brings significant stochastic disturbances to the power grids. It is worth of further discussion and research to find out whether the stochastic disturbances can induce new oscillation phenomena in the power grids [1–3].

Researchers have provided reasonable explanations based on theoretical analysis and oscillation reappearance for the oscillation generation mechanisms in the actual

P. Ju, *Stochastic Dynamics of Power Systems*, Power Systems,
https://doi.org/10.1007/978-981-13-1816-0_5

power grids. Nowadays, negative damping and forced power oscillation are the well-known and acceptable oscillation mechanisms in power grids. Power grids, along with increasingly rational structures and PSS devices in use, are able to provide enough damping during normal operation. Thus less and less oscillations are caused by the negative damping mechanism in power grids. By contrast, the forced power oscillation is a power grid oscillation phenomenon induced by a disturbance source. It may still occur under great damping conditions and may affect the stability of operations of the power grid. The forced oscillation mechanism indicates that forced oscillation in power system can be induced by continuing and periodic small disturbance. When the disturbance frequency is equal or close to the system's natural oscillation frequency, the system resonance will be induced, resulting in substantial oscillations or oscillations called the resonance mechanism [4, 5]. The forced oscillation mechanism not only provides ideas for traditional analysis on the oscillation mechanism, but also gives approaches for taking reasonable measures to suppress oscillations. However, this theory, mainly based on the deterministic input disturbance, draws a conclusion through a point-to-point corresponding relationship between the disturbance frequency and the system oscillation mode frequency [6–8]. This setting is less likely to cause oscillations and does not provide a good explanation for some recent oscillations that have taken place frequently. Besides, the theory is limited because it does not consider increasing stochastic factors in power grids. Above all, the traditional forced oscillations can be referred to as "special forced oscillations (SFO)".

The linear analysis method is mainly adopted for low-frequency oscillation in power system, such as the linearized system frequency-domain transfer model as well as for assuming the complete decoupling between system oscillation modes, etc. An actual power system can be analyzed through a linear system theory when it does not behave with strong nonlinear properties. However, the system operating status will gradually move away from its stable equilibrium point because of system faults and the increase of external disturbances, so that it shows strong nonlinearity. At this time, the oscillation modes of power system cannot be completely decoupled, but are interconnected in a certain way [9, 10]. These coupling effects are real and to some extent influence the dynamic response behavior of the system. Considering the nonlinear coupling between oscillation modes, the linearization method is subject to its limitations, which leads to discrepancy between analysis results and the actual situation. In severe cases, it may mislead the researchers' judgment of oscillation mechanisms, and result in a wrong choice of suppression control strategy. Hence, it is of practical significance to study how the power system nonlinear properties influence its dynamic response behavior.

The internal resonance phenomenon is highly important, especially to the nonlinear system dynamics, but it is uncommon in the power system dynamic behavior analysis. This is mainly due to rigorous oscillatory conditions of the internal resonance phenomenon. For instance, it requires a strict integer multiple between inherent frequencies of the system. So far, no internal resonance phenomenon has been reported occurring on the actual power grids. According to the study of general forced oscillations (GFO), the probability of power system oscillations caused by the stochastic disturbance is obviously larger than that of SFO. As a result, it is worth further researching whether the start-oscillating conditions, inducing probability and effect factors of power system resonance phenomena change or not, if under the condition of GFO (i.e. under the stochastic disturbance).

5.2 General Forced Oscillations under Small Stochastic Disturbance

5.2.1 Mechanism Analysis of General Forced Oscillation

The time-domain model has been adopted in most traditional power system oscillation analyses on account of its convenience for computer program implementation. But considering the stochastic factors in the power system, the stochastic rules and properties can be described more clearly by a frequency domain curve obtained from the frequency domain model. Therefore, this chapter adopts the frequency domain to describe dynamic response properties of the power system under the stochastic disturbance. Due to relatively small stochastic disturbance intensity in the power system at present, it can be approximately described by the linearized power system model.

Given the frequency domain transfer function in a power system as $H(f)$, the power spectral density (PSD) of the input stationary stochastic process u as $S_u(f)$, the PSD of the output stationary stochastic process y as $S_y(f)$, based on the linear system theory:

$$S_y(f) = |H(f)|^2 S_u(f) \tag{5.2.1}$$

Therefore, at a frequency or frequency band, only when $|H(f)|^2$ and $S_u(f)$ are both large, $S_y(f)$ would be large; when either $|H(f)|^2$ or $S_u(f)$ is 0 or very small, $S_y(f)$ would be 0 or very small. The following is to analyze the circumstances under which $|H(f)|^2$ will become large.

Given the Laplace transfer function corresponding to $H(f)$ as $G(s)$, its n poles as λ_i, m zeros as β_j, then:

$$G(s) = \frac{C(s)}{D(s)} = \frac{\alpha \prod_{j=1}^{m} (s - \beta_j)}{\prod_{i=1}^{n} (s - \lambda_i)} \tag{5.2.2}$$

Given that an oscillation mode corresponds to a conjugate complex root:

$$\lambda_i = -\sigma_i \pm j2\pi f_{oi} \tag{5.2.3}$$

The corresponding parts in the transfer function are:

$$Z_{oi}(s) = \frac{1}{[s-(-\sigma_i + j2\pi f_{oi})][s-(-\sigma_i - j2\pi f_{oi})]} \tag{5.2.4}$$

$$\begin{aligned} H_{oi}(f) = Z_{oi}(j2\pi f) &= \frac{1}{(j2\pi f + \sigma_i - j2\pi f_{oi})(j2\pi f + \sigma_i + j2\pi f_{oi})} \\ &= \frac{1}{4\pi^2(f_{oi}^2 - f^2) + \sigma_i(\sigma_i + j4\pi f)} \end{aligned} \tag{5.2.5}$$

If $f = f_{oi}$, then:

$$|H_{oi}(f_{oi})|^2 = \frac{1}{\sigma_i^2(\sigma_i^2 + 16\pi^2 f_{oi}^2)} \tag{5.2.6}$$

If this mode has a weak damping, i.e. when σ_i is very small, then $|H_{oi}(f_{oi})|^2$ is very large. Taking Mode 9 (its mode frequency is 0.57 Hz and a damping ratio is 6.26%) in the simulation system of this chapter as an example, its $|H_{oi}(f)|^2$ is shown in Fig. 5.1.

Thus, $|H_{oi}(f)|^2$ shows a sharp peak near the weak damping mode frequency f_{oi}. That is, near f_{oi}, $|H_{oi}(f)|^2$ is very large, but in other frequency bands far away from f_{oi}, $|H_{oi}(f)|^2$, it is very small.

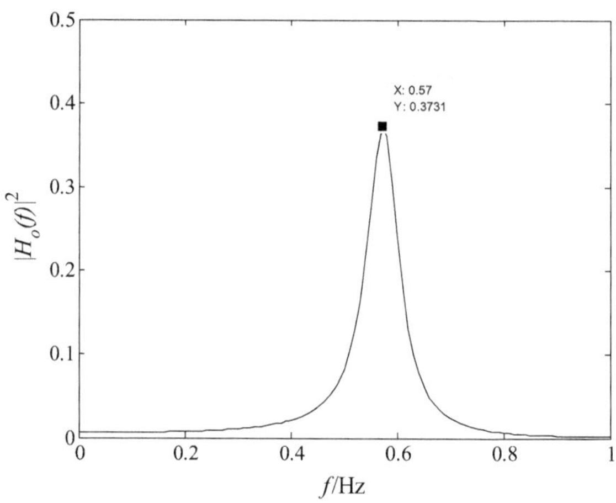

Fig. 5.1 Squared amplitude-frequency properties of a mode

If there are I such modes, then the amplitude-frequency properties corresponding to the I modes present a product relationship:

$$|H(f)|^2 = |C|^2 \prod_{i=1}^{I} |H_{oi}(f)|^2 \tag{5.2.7}$$

In other words, each amplitude-frequency property shows a sharp wave near each of the I mode frequencies. They are represented by a solid curve in Fig. 5.2.

Fig. 5.2 GFO mechanism diagram

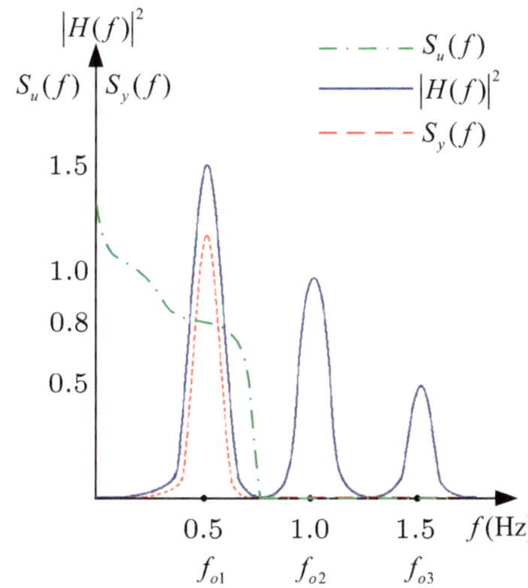

(a) Input-output schematic diagram 1

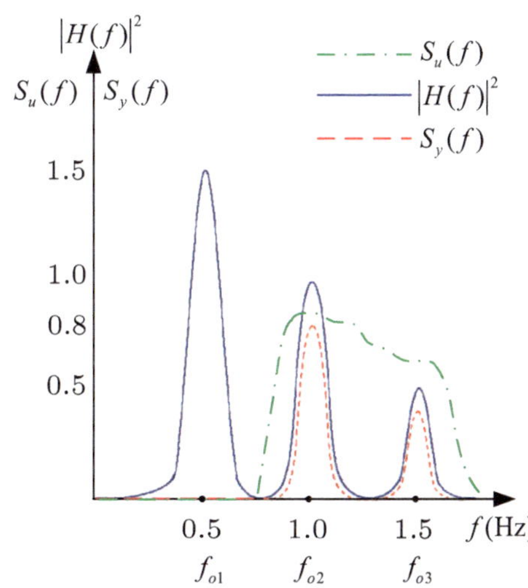

(b) Input-output schematic diagram 2

Assume that a certain input stochastic disturbance has a high power spectral amplitude at frequency bands corresponding to a weak damping mode frequency, as represented by a chain-dotted curve in Fig. 5.2a. Based on Formula 5.2.1, the output variable power spectral amplitude (represented by a dotted curve) obtained after multiplying becomes large, such as the weak damping mode f_{o1} in Fig. 5.2a. Instead, at a frequency band corresponding to a weak damping mode frequency, the input stochastic disturbance power spectral amplitude is small, such as the weak damping modes f_{o2} and f_{o3} in Fig. 5.2a. However, if the power spectral frequency band of the input stochastic disturbance changes, as shown in Fig. 5.2b, the output variable power spectral amplitude in the weak damping mode is small at f_{o1} and becomes large at f_{o2} and f_{o3}. Hence, the generation mechanism of power system oscillations caused by stochastic disturbances can be explained as follows: if a power spectrum frequency band of the input stochastic disturbance can cover a weak damping mode frequency, a large oscillation may be induced. The oscillation includes all frequency components of these weak damping modes.

It should be noted that the generation mechanism of SFO can also be explained by the theory above. If a sinusoidal disturbance is input near a certain weak damping mode frequency, its power spectrum is a vertical line, as shown by the chain dotted line in Fig. 5.3a. According to the formula (5.2.1), the output variable power spectrum after multiplying is also a vertical line (as shown by the dotted line). In other words, the output variables share a same sinusoidal oscillation curve with the disturbance frequency. Moreover, the closer to the weak damping mode frequency f_{o3} the disturbance frequency is, the larger the output power spectral amplitude becomes. On the other hand, if the disturbance frequency shows obvious deviation from the weak damping mode frequency, the output variable power spectral amplitude becomes small or even 0, as shown in Fig. 5.3b.

It should be specially pointed out that the condition of SFO is under the assumption that an oscillation source is a single frequency and equal to or approximately equal to a natural oscillation frequency, which rarely happens. Under the conditions found in this section, that is inducing oscillations under stochastic disturbance, a stochastic disturbance power spectral band capable of covering a weak damping mode frequency would be enough to significantly increase the oscillation probability. It may explain why some oscillations in the power system at present become forced oscillations with no reason found. It is also possible to explain why the power system oscillations have increased recently. The oscillations caused by the stochastic disturbance are forced oscillations, without being limited to SFO conditions. Therefore, the oscillation induced by the stochastic disturbance is called the "general forced oscillation (GFO)" in this section, while the SFO can be regarded as a special case of GFO.

Based on the above mechanistic analysis, a mathematical definition of GFO can be given as follows:

Fig. 5.3 SFO mechanism diagram

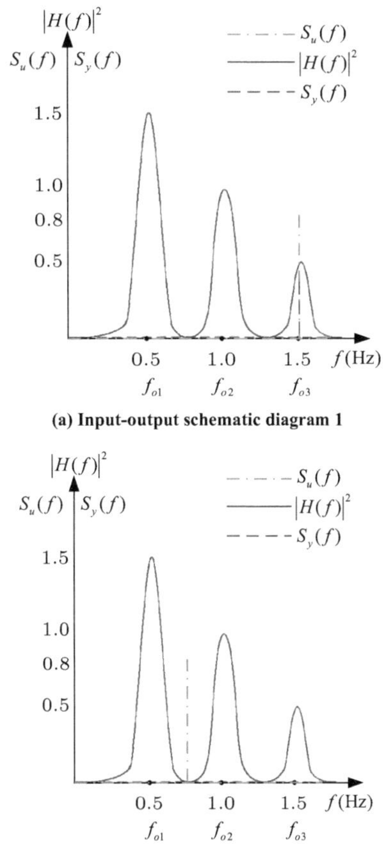

(a) Input-output schematic diagram 1

(b) Input-output schematic diagram 2

If the stochastic disturbance power spectral energy is mainly concentrated in a certain range of frequency band $[f_m, f_n]$, its power spectrum can be described as:

$$S_u(f) = \begin{cases} g(f), & f \in [f_m, f_n] \\ \varepsilon(f), & f \notin [f_m, f_n] \end{cases} \tag{5.2.8}$$

Wherein $g(f)$ and $\varepsilon(f)$ respectively represent amplitude-frequency properties of the stochastic disturbance in different frequency ranges, which satisfy $\bar{\varepsilon} << \bar{g}$. If some weak damping mode frequencies in the power system satisfy $f_i \in [f_m, f_n]$, an oscillation, including all the weak damping mode frequency components, is induced with a high amplitude. This type of oscillation is defined as GFO. Particularly, if $f_m = f_n$, the disturbance becomes a periodic disturbance. If some weak damping mode frequencies in the power system satisfy $f_m = f_n = f_i$, a periodic oscillation, with an oscillation frequency equal to the disturbance frequency, is induced. This type of oscillation is defined as a traditional SFO.

The GFO theory can be used to explain the generation mechanism of single machine infinite bus oscillations induced by the ideal Gaussian white noise disturbance. The Gaussian white noise is a stochastic process during which a power spectrum is uniformly distributed in an entire frequency domain. That is, it has the same energy density at each frequency point and can cover the inherent frequency of a single machine infinite bus system, resulting in the occurrence of GFO.

5.2.2 Simulation Analysis of General Forced Oscillation

A 10-generator and 39-bus New England Test System is used to verify the analysis, as shown in Fig. 5.4. Note that G10 is an external equivalent generator which has a large inertia time constant. Therefore, it is selected as the reference generator. A 5-order detailed model is adopted for generators, while a constant impedance model is used for loads. By using small signal stability analysis modules in PSASP, 9 electromechanical oscillation modes and major participating generators thereof can be obtained, as shown in Tables 5.1 and 5.2. As shown in Table 5.1, the frequencies of the 9 modes are in the range of low-frequency oscillation. In

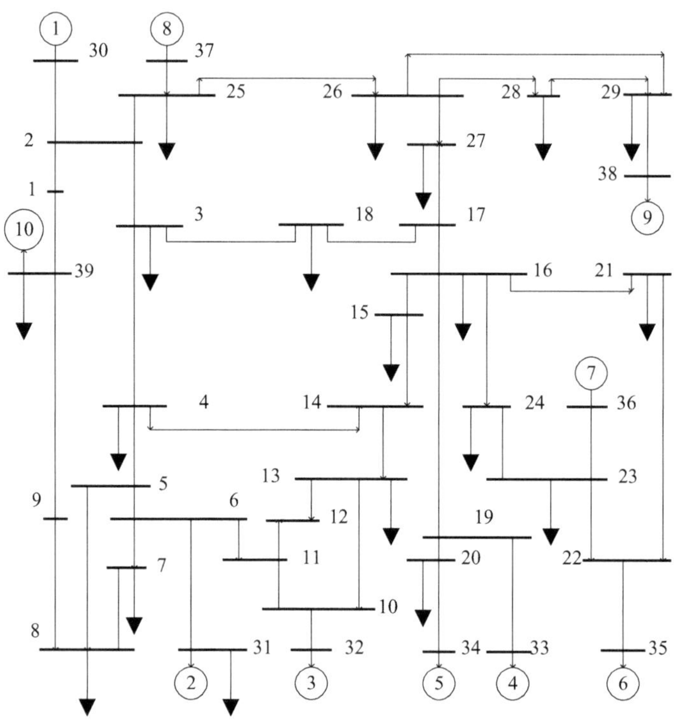

Fig. 5.4 Simulation system diagram

Table 5.1 System oscillation modes

Oscillation mode	Frequency (Hz)	Damping ratio (%)
1	1.466892	6.0419
2	1.456350	5.2050
3	1.447361	4.1513
4	1.201765	3.4492
5	1.191107	4.9570
6	1.137658	4.8310
7	1.000131	4.2776
8	0.904576	4.5232
9	0.573959	6.2630

Table 5.2 Major participant generators in system oscillation modes

Oscillation modes	Participant generators
1	G4<->G5
2	G7<->G6
3	G8<->G1
4	G1, G8<-> G9, G3, G6, G7
5	G2<->G3
6	G5, G4<->G6, G7
7	G2, G3<->G5, G6, G4, G7
8	G9<->G5, G6, G3, G4, G7, G2
9	G10<->G9, G6, G5, G4, G7, G3, G2, G1, G8

addition, the participation degree of each generator in each oscillation mode decreases progressively from left to right at both ends of the symbol "<->", while the generators with small participation factors are not listed in.

As shown in the modal analysis Table 5.2, generator G5 is involved in 5 electromechanical oscillation modes, which are Mode 1, 5, 7, 8, and 9 respectively. Their oscillation frequencies are in the range of 0.5–1.5 Hz. Besides, it participates in the 5 modes to a relatively high degree, thus facilitating the contrastive analysis. Thus, the bus 34 connected to G5 is selected as the equivalent current injection point, in order to simulate the stochastic fluctuations of generator active powers. In the simulation, the active power of G5 is 5.08 p.u. (508 MW). According to the GFO theory, if the stochastic disturbance power spectral frequency band can cover an oscillation mode frequency in the system, the oscillation under this mode frequency will be induced. To verify the theory, firstly, assume that the power spectral frequency band of the stochastic disturbance is 0–3 Hz, while the amplitudes of the remaining frequency bands are 0. Thus, 5 oscillation mode frequencies where G5 participates are covered. Then, based on a frequency-time conversion method, time-domain sequences of stochastic disturbance can be obtained, as shown in Fig. 5.5. The power in this figure is a per-unit value, the same as below.

Fig. 5.5 Power spectrum and time-domain sequences of stochastic disturbance

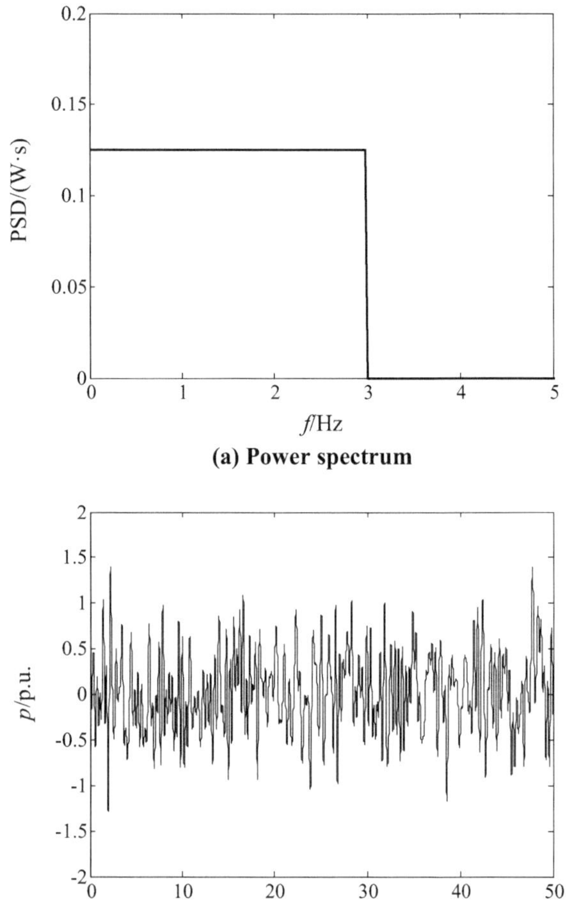

(a) Power spectrum

(b) Time-domain sequences

Apart from that, the stochastic disturbance time-domain sequences after the time-frequency conversion are injected into the simulation system. The relative power angle of G5 against G10 and the active power of the inter-area tie-line located between buses 16 and 19 are chosen as the output variable. After obtaining the time-domain curve, the power spectrum analysis is conducted, with results shown in Figs. 5.6 and 5.7. As shown, the stochastic disturbance can induce a forced oscillation phenomenon. It can be seen that according to a power spectral peak frequency marked in Fig. 5.7, when a stochastic disturbance power spectrum frequency band covers 5 oscillation mode frequencies in which G5 is involved, 5 peaks of corresponding frequencies can be observed from the G5 relative power angle power spectrum. Based on the active power results of the inter-area tie-line in Fig. 5.7, 4 peaks can be observed but they do not include Mode 1. It can be seen in

Fig. 5.6 G5 relative power angle under wide band stochastic disturbance

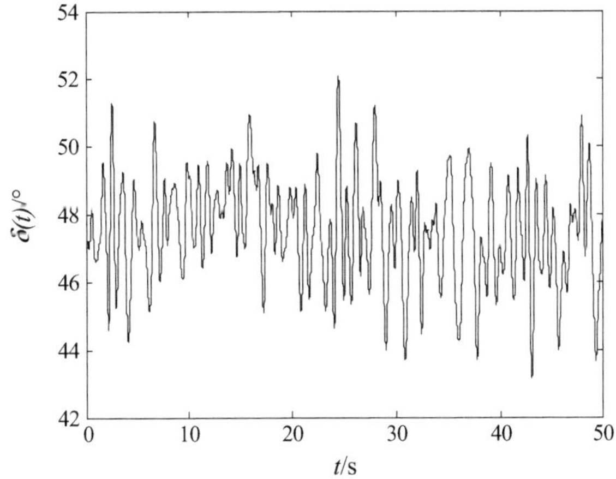

(a) Relative power angle

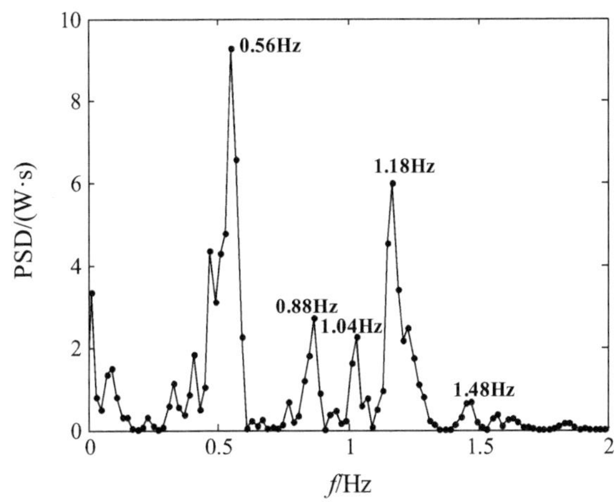

(b) Power spectrum

the analysis that Mode 1 is a local oscillation mode between G4 and G5 and that the inter-area tie-line lies outside G4 and G5, so that Mode 1 is out of view in the tie-line power spectrum.

It should be noted that due to the non-linearity of the simulation test system and due to errors in numerical calculation, the peak frequencies of the power spectrum in Figs. 5.6 and 5.7 have slightly deviated from those of the system oscillation mode. Through the verification by calculating the relative power angle power spectrum of each unit, it can be concluded that the 5 oscillation modes that G5 is involved in are all excited, but none of the 4 oscillation modes that G5 is not involved in are excited.

Fig. 5.7 Inter-area tie-line active power angle under wide band stochastic disturbance

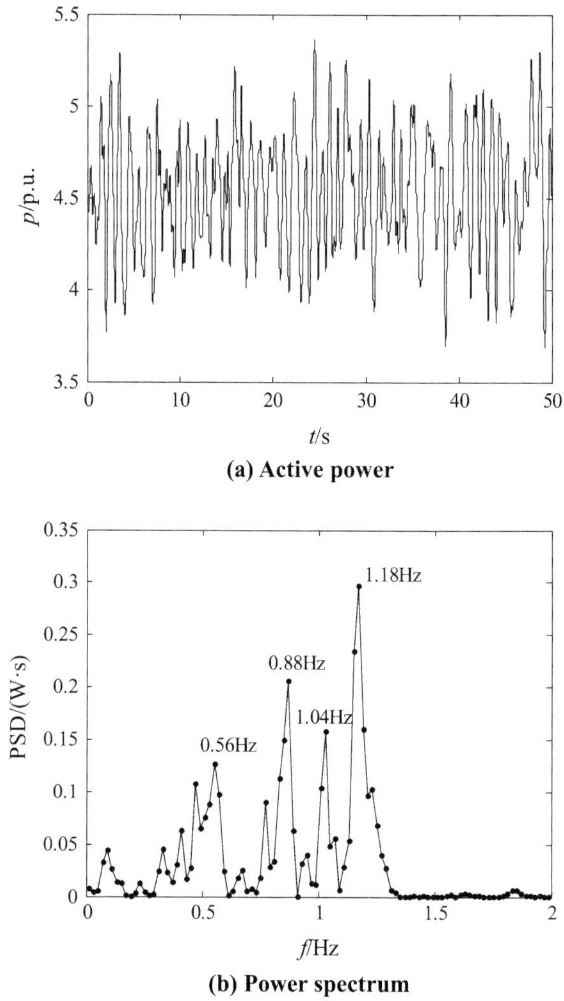

To further validate the general forced oscillation theory, a full-band stochastic disturbance is divided into two narrow frequency bands and selectively excited for the 5 oscillation modes in which G5 participates. Assume that the power spectral amplitude under the original stochastic disturbance does not change in the two frequency bands 0.4–0.7 Hz and 1.1–1.4 Hz. The amplitudes of the remaining frequency bands become zero, as shown in Fig. 5.8. In order to fit in the stochastic power fluctuations that appeared in the actual power grid, the injection point of low-frequency stochastic disturbance is selected at bus 34 connected to G5 to simulate the slow-varying stochastic fluctuation of a generator active power. Instead, the injection point of high-frequency stochastic disturbance is selected at load bus 20, which has the closest electrical distance to G5, to simulate the

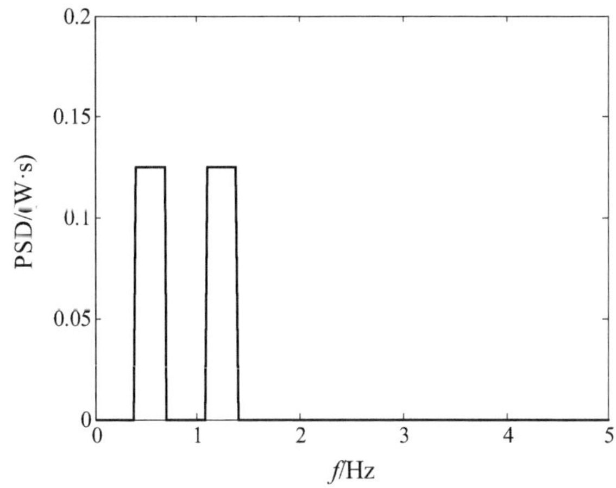

Fig. 5.8 Power spectrum under two-band stochastic disturbance

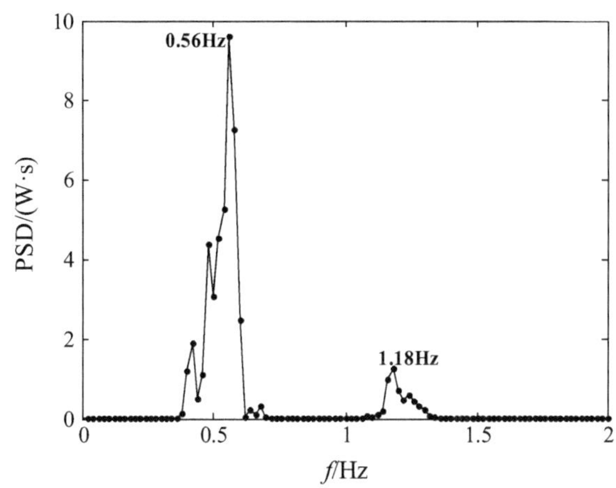

Fig. 5.9 G5 relative power angle power spectrum under two-band stochastic disturbance

fast-varying stochastic fluctuation of the load. The power spectrum of the G5 relative power angle obtained under the stochastic disturbance of the generator power and load power is shown in Fig. 5.9. The power spectrum of tie-line 16–19 is shown in Fig. 5.10.

It can be seen from the results of modal analysis that the power spectrum of the two-band stochastic disturbance only covers frequencies corresponding to Mode 6 and Mode 9, which G5 participates in. From the simulation results in Figs. 5.9 and 5.10, only two peaks are observed, of which the peak frequencies respectively correspond to Mode 6 and Mode 9. It can be concluded that an oscillation mode cannot be excited when the power spectrum of stochastic disturbance fails to cover the frequency corresponding to this mode. This further verifies the correctness of the GFO theory. In addition, it can be found from the comparison between Figs. 5.9 and 5.10 that not only the generator active power stochastic fluctuation can induce a

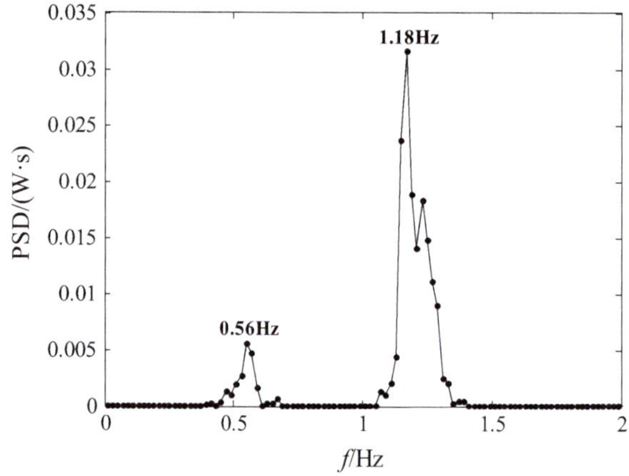

Fig. 5.10 Inter-area tie-line active power spectrum under two-band stochastic disturbance

system forced oscillation, but also the load power stochastic fluctuation electrically close to the generator can impact the system oscillation. And the closer a disturbance source is to an observation point in electrical means, the greater the impact becomes.

The comparative analysis between GFO and SFO is provided as follows. First, effects of an oscillation mode frequency on the GFO amplitude are analyzed. According to Table 5.1, the damping ratios of Mode 1 and Mode 9 are relatively close, but their natural oscillation frequencies are quite different. The corresponding amplitude-frequency properties of both modes are shown in Fig. 5.11. It can be obviously seen that when damping ratios of the two oscillation modes are close, the amplitude-frequency properties at different frequencies differ distinctly. Besides, amplitude-frequency properties are directly related to the system output power spectrum amplitude. Therefore, the influence of the mode frequency on GFO should be noted.

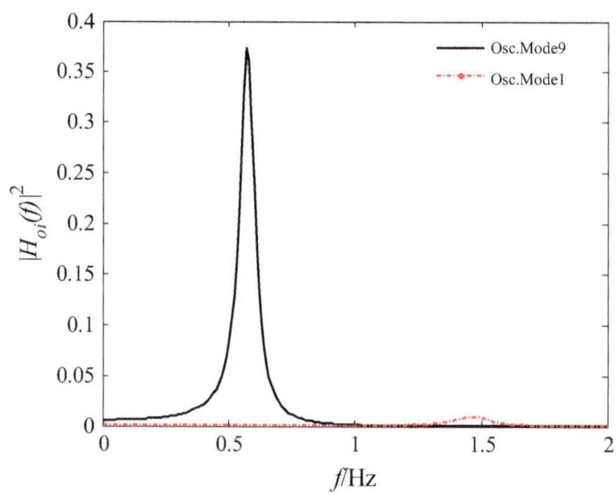

Fig. 5.11 Comparison of amplitude-frequency properties between two modes

Fig. 5.12 Power spectrum and time-domain sequences of stochastic disturbance

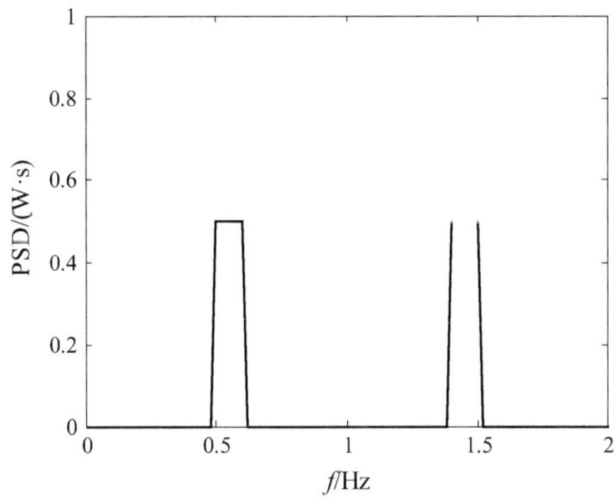

(a) Power spectrum

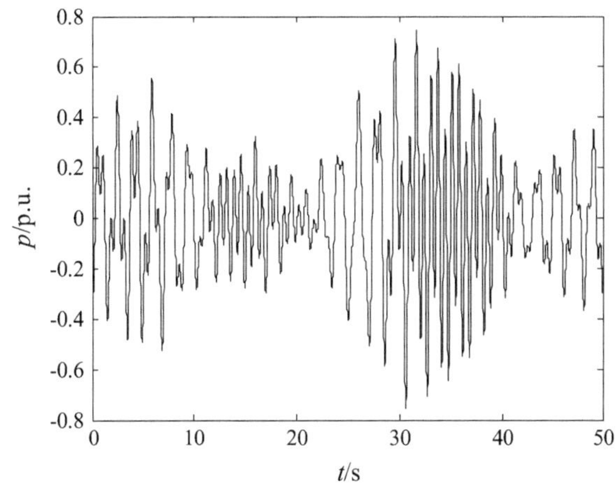

(b) Time-domain sequences

As seen in Table 5.2, G5 is involved in both oscillation modes concurrently. Thus, it should be considered to select bus 34 connected to G5 as the equivalent current injection point to simulate the stochastic fluctuation of the generator's active power. To excite forced oscillations of the two oscillation mode frequencies and exclude the influence of other modes, it can be assumed that the stochastic disturbance power spectral frequency bands range between 0.5 and 0.6 Hz (Coverage mode 9 frequency) and between 1.4 and 1.5 Hz (Coverage mode 1 frequency). Then, based on the frequency-time conversion method, the time-domain sequences of stochastic disturbance are obtained as shown in Fig. 5.12.

Eventually, the stochastic disturbance time-domain sequences after the time-frequency conversion are injected in the simulation system. The relative power

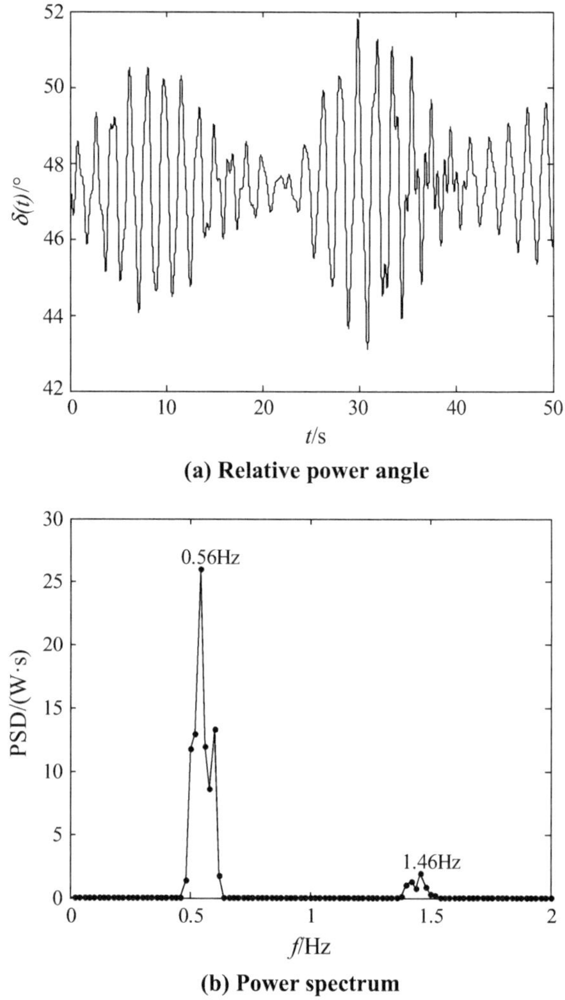

Fig. 5.13 Power spectrum and relative power angle of G5 under two-band stochastic disturbance

angle of G5 against G10 is chosen as the output variable. After obtaining the time-domain curve, the power spectrum analysis is conducted, with results shown in Fig. 5.13. According to Fig. 5.13, as the frequency band of the input disturbance power spectrum covers the natural oscillation frequencies of Mode 1 and Mode 9, the oscillation of two mode frequencies is excited. In addition, the oscillation amplitude corresponding to the frequency of Mode 9 is greater than that of Mode 1. Combining this with the analysis in Fig. 5.11, it can be summarized that under the same disturbance intensity, if the oscillation at frequencies of a number of weak damping modes is excited, the amplitudes of oscillation components at lower frequencies are higher, causing greater harm to the system.

It should be noted that the input and output locations are the same. Thus, the influence of electrical distance between input and output on oscillation amplitudes is omitted and only the influence of oscillation mode is analyzed. In fact, the electrical distance is another vital factor that influences a response oscillation amplitude. This factor is calculated and analyzed as follows.

According to Fig. 5.4, load buses 29, 28, 26, and 27 share a close electrical distance with G9. Therefore, it is considered that these 4 buses can be selected as equivalent current injection points to simulate the stochastic fluctuations of loads at different locations. Then, to make the electrical distance between input and output more visualized and easier to compare, output observation points are located in G9, which can also avoid influence on other generators. The relative power angle of G9 against G10 is selected as the output variable. Finally, it is assumed that the frequency band of the stochastic disturbance power spectrum falls in 0.8–1.0 Hz to excite the oscillation Mode 8 in which G9 participates to a high degree, as shown in Fig. 5.14.

Figure 5.15 shows the power spectrum analysis results of relative power angles at different disturbance locations, i.e., bus 29 and 28, are shown respectively. It can be seen that under the same stochastic disturbance with different electrical distances between the disturbance source and observation points, the output response power spectrum shows similar status, but significant difference in amplitudes. Given that the stochastic disturbance remains the same, an equivalent current is injected in the other 2 load buses respectively. The peak amplitudes of the relative power angle power spectrum at G9 are compared, as shown in Table 5.3.

According to Table 5.3, under the same stochastic disturbance, because of the expanded electrical distance between the disturbance source location and G9, the power spectral amplitude of the relative power angle at G9 continues to decrease. It can be concluded that the electrical distance between an input disturbance and an output response can significantly affect the oscillation amplitude. The smaller the electrical distance is, the more drastic the induced oscillation would be.

Fig. 5.14 Power spectrum of stochastic disturbance

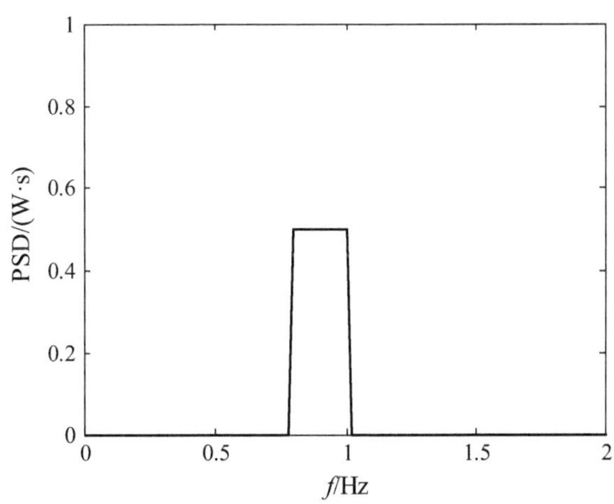

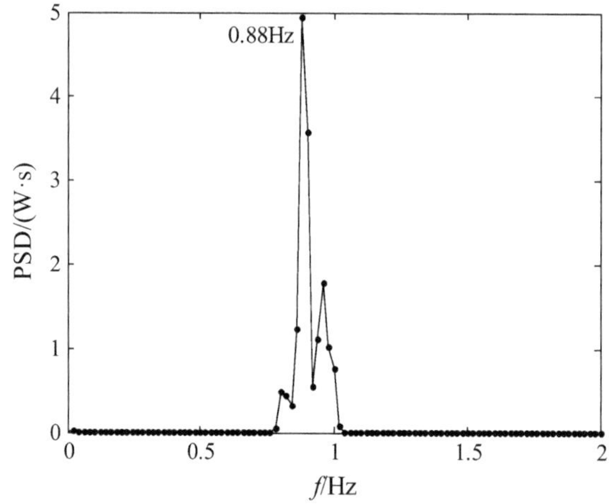

(a) Disturbance source location: bus 29.

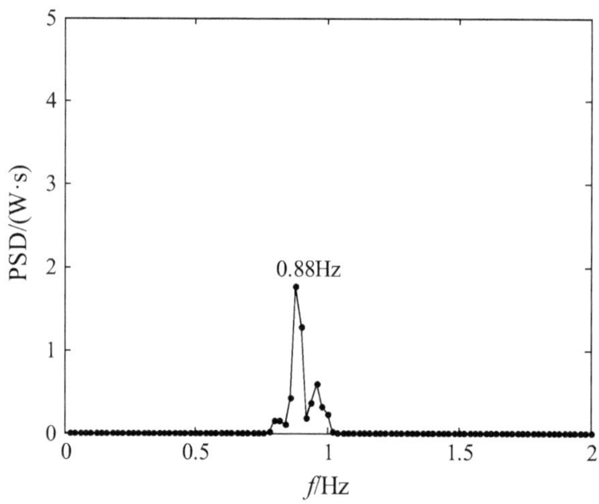

(b) Disturbance source location: bus 28.

Fig. 5.15 Relative power angle of G9 under stochastic disturbance

Table 5.3 Amplitudes of output power spectrum

Disturbance source location	Amplitudes of output power spectrum
Bus 29	4.953
Bus 28	1.768
Bus 26	1.752
Bus 27	1.675

Table 5.4 Comparison and summary

Aspects	Special forced oscillations	General forced oscillations
Disturbance source available or not	Yes	Yes
Disturbance source frequency	Single-frequency	Frequency-band
Start-oscillating condition	Disturbance frequencies are equal or close to the natural oscillation frequency	Disturbance frequencies cover the natural oscillation frequencies
Output response	Single-frequency oscillation	Stochastic multi-frequency oscillation
Main influencing factors	Input disturbance intensity and location, damping ratio, system parameters, system state	Input disturbance intensity and location, damping ratio, mode frequency, system parameters, system state
Occurrence probability	Point-to-point, small	Plane-to-point, high

In conclusion, a comparison between GFO and SFO is summarized and listed in Table 5.4. The key points of this comparison are as follows. (1) GFO is a breakthrough and extension of SFO, and SFO is a special example of GFO. (2) Compared with the "point-to-point" start-oscillating condition of SFO, the "plane-to-point" start-oscillating condition of GFO significantly increases the possibility of oscillation in power system. This may explain the increasing power system oscillation in recent times. (3) Among the influencing factors of the forced oscillation, the electrical distance between input and output is of remarkable importance. The smaller the electrical distance is, the more drastic the oscillation would be. (4) Among the influencing factors of GFO, the effect of oscillation mode frequency needs to be considered. The lower the frequency is, the larger the oscillation component amplitude would be.

5.2.3 *Practical Analysis of General Forced Oscillation*

5.2.3.1 Measurement and Analysis of Inter-area Oscillation in Henan Power Grid

(1) **Introduction of the Henan Power Grid**

The Henan Power Grid is located at the electrical intersection of the power grids in Northwest, North China and Central China, as shown in Fig. 5.16. As is seen in the alternating current system, Henan Power Grid, together with the North China Power Grid in the north and the Central China Power Grid (four provinces and one city in the south) has formed a three generator groups and two AC transmission sections

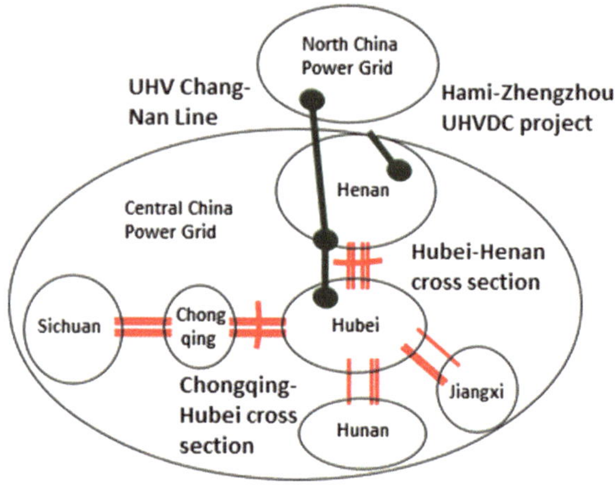

Fig. 5.16 General distribution of Henan Power Grid

system. From the UHV level, the Henan Power Grid has formed a typical receiving system of UHV AC-DC hybrid power grids. As the above said power grids become more interconnected, the Henan Power Grid has proven to be an important hub to balance supply and demand in power grids of all these regions.

Limited by monitoring and analysis techniques for a long time, oscillation phenomena in Henan Power Grid are difficult to record, not to mention statistical analysis and off-line data quantization modeling. No analysis in detail or effective troubleshooting has succeeded in addressing oscillation problems so far. To a certain extent, this restricts further improvement of the power grid transmission capacity. In recent years, with the ever-improving D5000 system of the State Grid Corporation of China, D5000/WAMS, at least in province level areas, has been utilized to detect and record anomalous fluctuations in millisecond level. It can perform statistics in different dimensions like frequency, oscillation times, amplitude, geological locations, and provide great fundamental technical support for analyzing and controlling power grid oscillations. Besides, it also furnishes numerous data records for the theory study of power grid oscillations.

(2) **Investigation of the oscillation phenomenon in Henan Power Grid**

With the rationalization of interconnected power grid structures and massive PSS installation, the stability of Henan Power Grid has been remarkably improved. In recent years, less widely-spread oscillation accidents with great perniciousness occurred. However, the load grows rapidly, there is large-scale interconnection between regions, and the operation mode flexibility is increased. Thus, the Henan Power Grid is still at risk of widely spread oscillation events.

Generally, the analysis of power system oscillations consists of the long-term steady-state fluctuation and electromechanical oscillations within 0.1–2.5 Hz. This chapter mainly analyzes electromechanical oscillation phenomena recorded by D5000/WAMS in the Henan Power Grid. Monthly statistics of daily power fluctuation times all over 2014 in the Henan Power Grid are shown in Table 5.5. The

Table 5.5 Monthly statistics of daily power fluctuations in 2014 in Henan Power Grid (times)

Month frequency (Hz)	1	2	3	4	5	6	7	8	9	10	11	12
0.15–0.20	0	0	0	0	0	111	621	653	702	483	386	437
0.30–0.45	6	17	2	57	24	129	177	214	350	717	1772	1276
0.75–0.90	1100	1472	75	30	158	92	21	237	731	2163	751	378
Over 1.0	52	708	1	109	41	9	15	7	17	79	123	404

partial results of oscillation fluctuation recording are given in Figs. 5.17, 5.18, 5.19 and 5.20 with the oscillation frequency, oscillation amplitude and duration that are noted. In terms of the oscillation frequency, there are mainly 4 oscillation modes in the Henan Power Grid. (1) Large inter-area oscillation mode at 0.15–0.20 Hz between the North China Power Grid and Central China Power Grid. (2) Large inter-area oscillation mode at 0.30–0.45 Hz between the Sichuan Power Grid and Central China Power Grid. (3) Small inter-area oscillation mode at 0.75–0.90 Hz between the North Henan Power Grid and the Central Henan Power Grid. (4) Local oscillation mode at over 1.0 Hz within part of the Henan Power Grid generators. Due to different daily operations, the oscillation frequency in a certain range will be different. But the oscillation modes which Henan Power Grid generators participate in are basically the same. In terms of the oscillation times, Mode 1 and Mode 2 are more frequent oscillation modes in Henan Power Grid, which averagely occurred for several or even dozens of times per day. This is worth of further analysis and study.

(3) **Oscillation Times Statistics and Power Spectral Analysis**

The distribution of daily oscillation times in oscillation Mode 1 and Mode 2 occurred in January, 2015 is respectively shown in Fig. 5.21. It can be seen that the oscillation times of Mode 1 are approximately uniformly distributed during January. However the oscillation times of Mode 2 are intensive at the middle of this month but less in remaining days. The whole times are irregularly distributed.

For system spontaneous oscillations caused by faults or special forced oscillations caused by disturbance sources, the time distribution of oscillation times is also stochastic due to uncertainties of the occurrence of faults or disturbance sources. The distribution is similar to that shown in Fig. 5.21b. In terms of GFO induced by stochastic disturbance from the renewable energy power generation or new-type load, continuous stochastic power oscillations should be frequently observed in actual power grid operations due to the persistent disturbance source. The distributions are similar to Fig. 5.21a. Thus, the power oscillation at Mode 1 frequency occurring in the power grid may be GFO induced by the stochastic disturbance. Since Mode 1 is a large inter-area oscillation mode between the North China Power Grid and the Central China Power Grid, it is most appropriate to analyze the measured power fluctuation data on 1000 kV UHV Chang-Nan regional tie line.

Based on the monitoring data of D5000/WAMS, it is found that there are daily power oscillations of about 400 MW on 1000 kV UHV Chang-Nan Line. In some special cases, the oscillation in Chang-Nan Line is excited up to 1000 MW. The

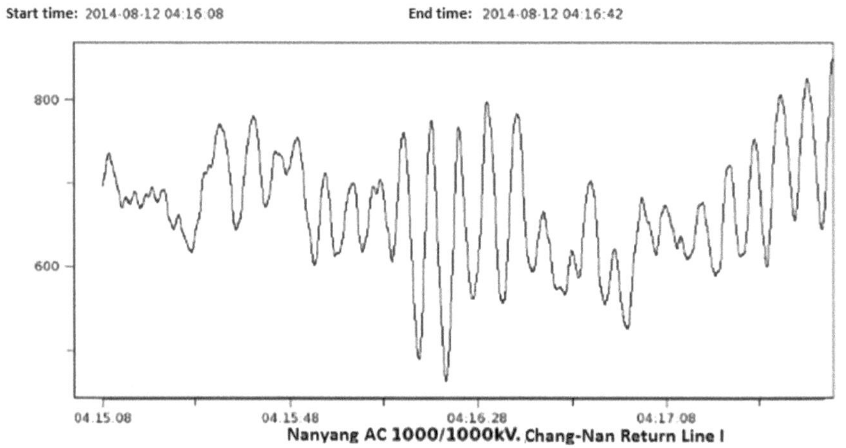

(a) 0.1750Hz, 248MW, 34s

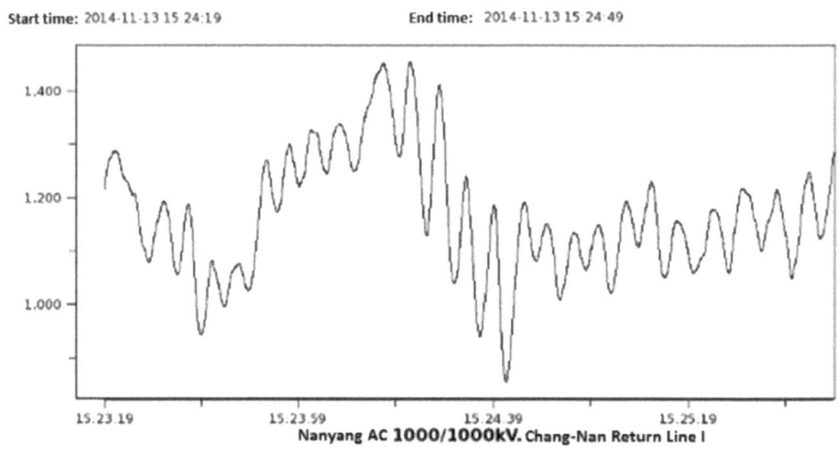

(b) 0.1790Hz, 286MW, 30s

Fig. 5.17 Central China-North China large inter-area oscillation mode in which Henan Power Grid participated

time-domain curves of power oscillation in Chang-Nan Line are shown from 04:00 to 05:00, from 10:00 to 11:00 and from 22:00 to 23:00 on January 1, 2015 respectively with the corresponding power spectral analysis results in Fig. 5.22. It can be seen that although the time-domain curves of power oscillations in Chang-Nan Line at different time periods are complicate, their frequency-domain

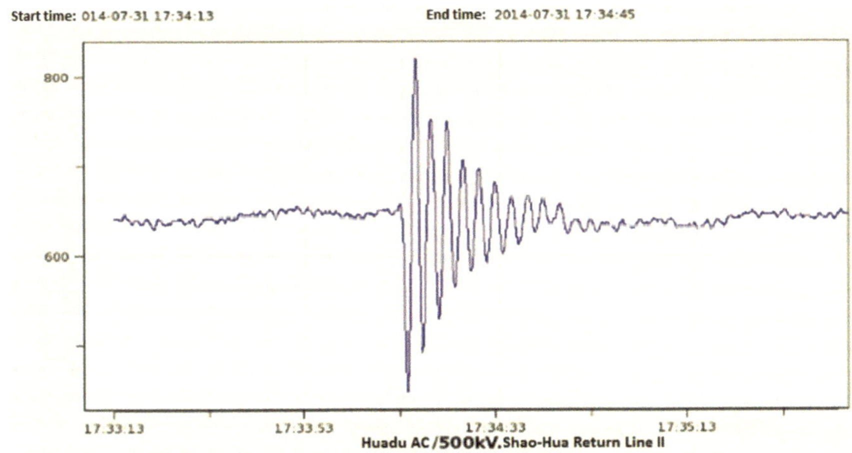

(a) 0.3086Hz, 132MW, 32s

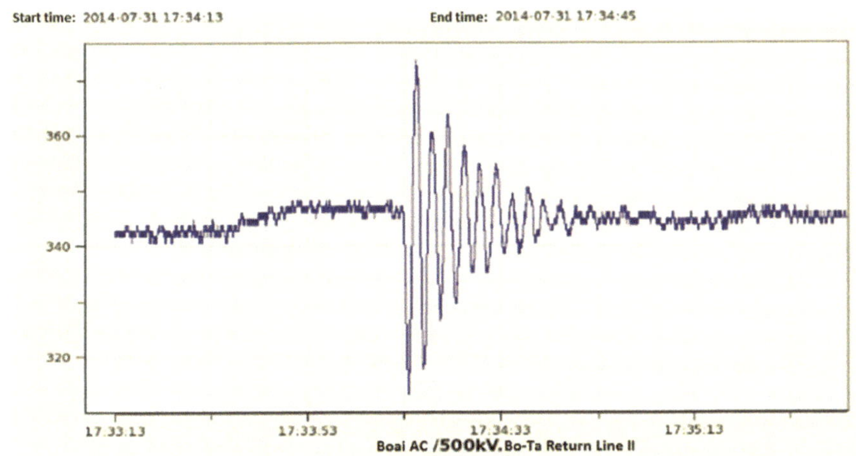

(b) 0.3086Hz, 132MW, 32s

Fig. 5.18 Sichuan-Central China large inter-area oscillation mode in which Henan Power Grid participated

properties are quite similar. The oscillation energy is mainly distributed in 0.1–0.25 Hz. But the oscillation energy is very small at remaining frequencies, and the maximum energy amplitude appears at about 0.17 Hz. These features coincide well with the GFO properties. That is, the oscillation energy is concentrated in a certain frequency band and extrema occur at the system oscillation mode frequencies.

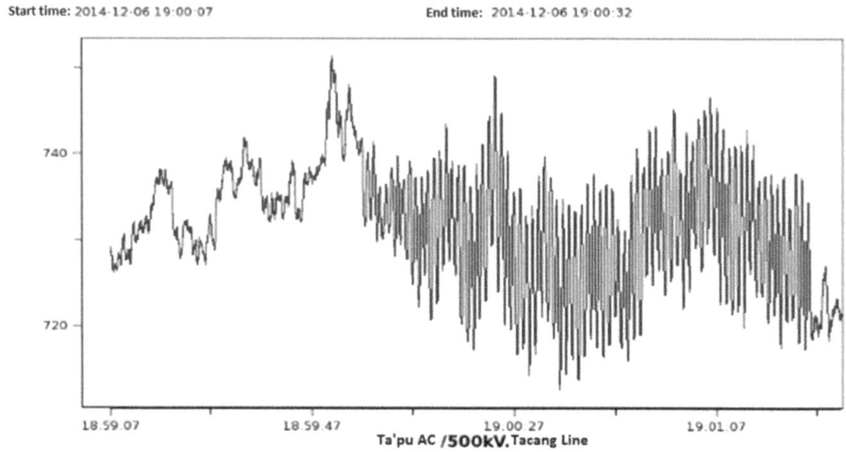

(a) 0.8491Hz, 15MW, 52s

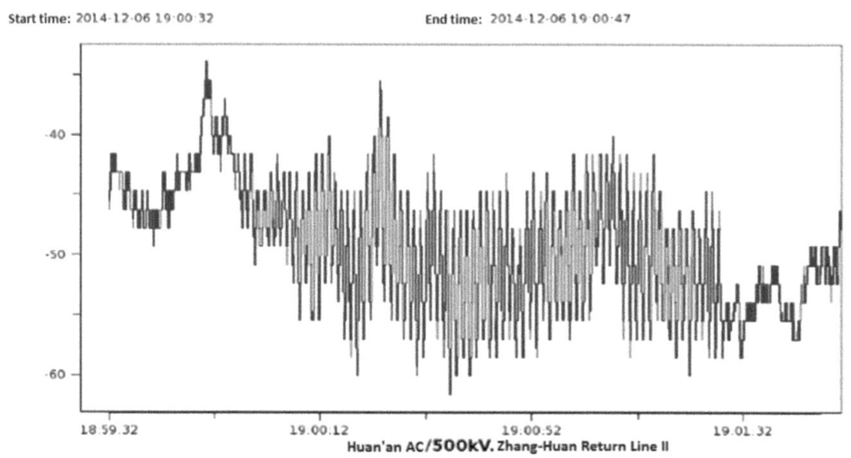

(b) 0.8491Hz, 15MW, 52s

Fig. 5.19 North Henan-Central Henan inter-area oscillation mode in Henan Power Grid

5.2.3.2 Measurement and Analysis of Wind Farm Power Fluctuation

In recent years, owing to technological improvement in the wind energy sector and the declining construction cost, wind power generation has become a new power generation method with the largest application scale, and has played an important role in China's power supply. However, the wind speed often features high volatility and intermittency, and is easy to be affected by geological locations and surrounding environment, such as the tower shadow effect and wake effect, etc.

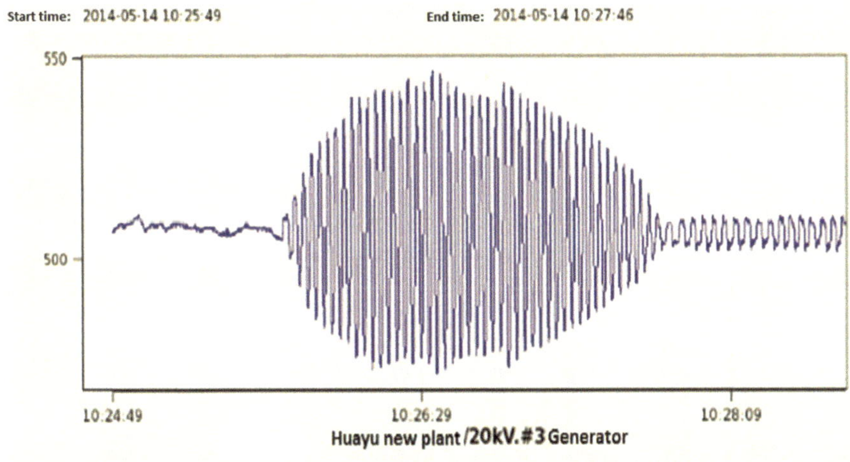

(a) 0.3886Hz, 9MW, 117s

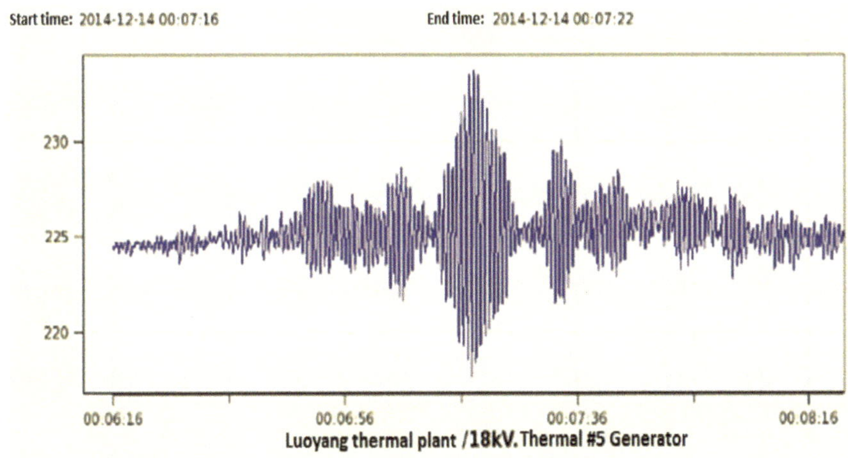

(b) 1.3554Hz, 13MW, 52s

Fig. 5.20 Local oscillation mode in part of generators in Henan Power Grid

This feature directly leads to a strong stochastic property of the power out of the wind power conversion, affecting the safety and stability of power grid operation [11–13].

By December 2015, the wind power installation capacity in Henan Power Grid had reached 300 MW. In order to study the influence of wind power stochastic fluctuation on the power system oscillation, it is necessary to summarize the time-frequency and frequency-domain features in the wind power fluctuation. Therefore, the wind power data from 04:00 to 05:00, from 10:00 to 11:00 and from

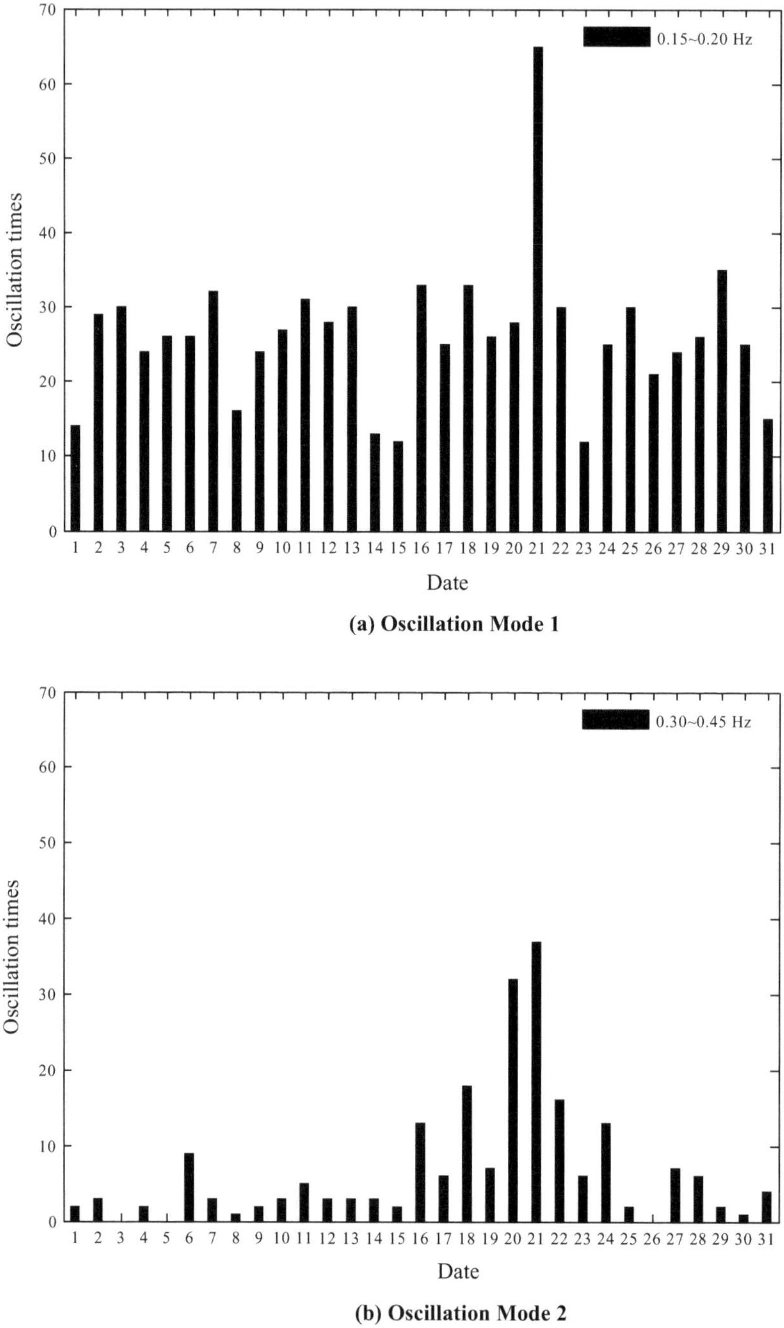

(a) Oscillation Mode 1

(b) Oscillation Mode 2

Fig. 5.21 Analysis of oscillations in two oscillation modes

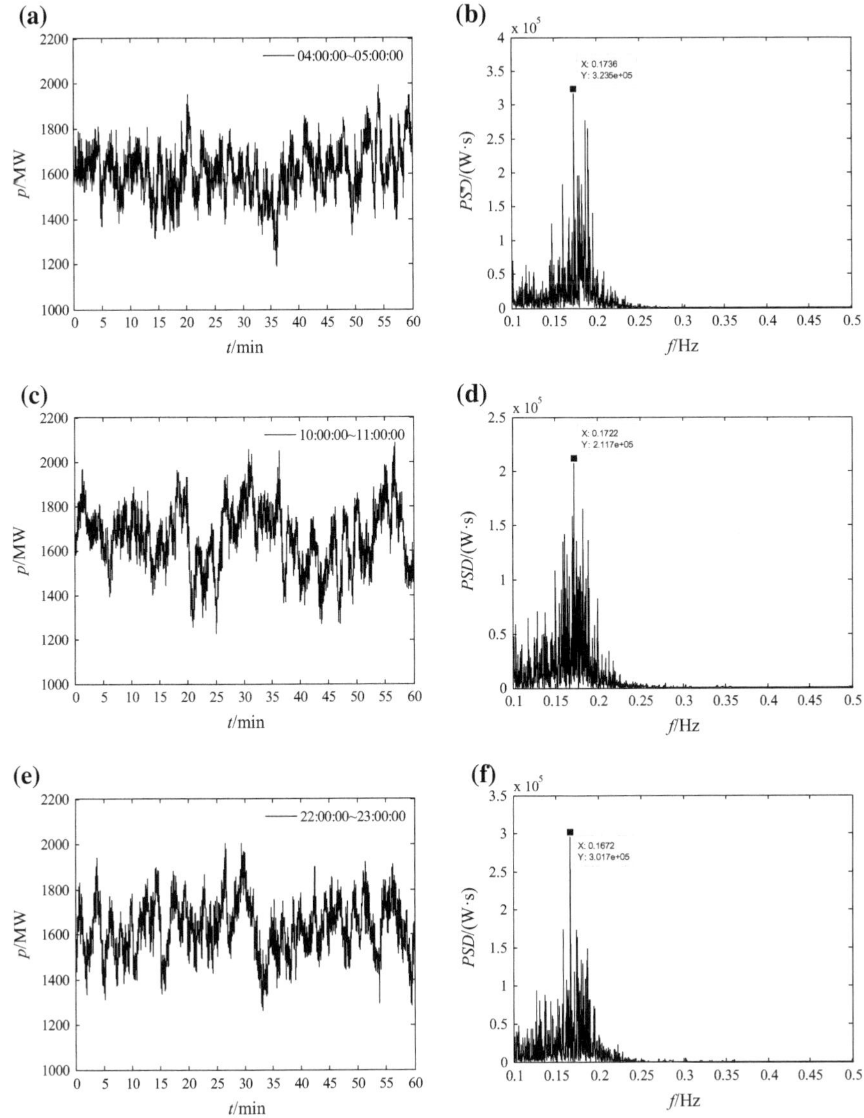

Fig. 5.22 Chang-Nan Line power fluctuation curves and their corresponding power spectrum in different time periods

22:00 to 23:00 are selected to perform power spectral analyses and to research their frequency-domain properties. Generally, the power system oscillation frequency is between 0.1 and 2.5 Hz. Thus, as shown in Fig. 5.23, this section mainly focuses on the power spectral properties of wind power fluctuation within this range. It can be seen in the graph that although the wind power performs differently in different time periods with highly stochastic power fluctuations, the frequency properties are

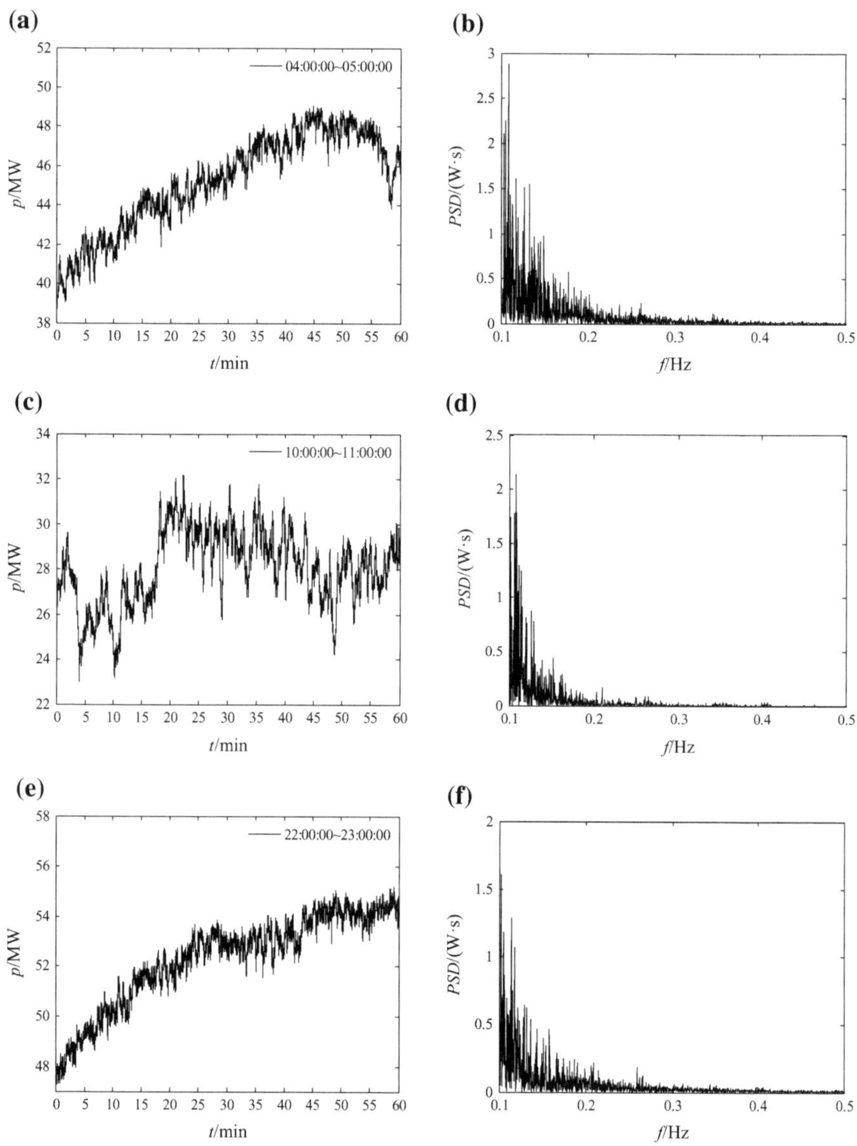

Fig. 5.23 Wind power fluctuation curves and the corresponding power spectrum in different time periods

rather similar. The distribution of fluctuation energy ranges generally between 0.1 and 0.3 Hz, while the energy levels are very low at other frequencies. In addition, the frequency band happens to cover the frequency of the large inter-area oscillation mode between the North China Power Grid—Central China Power Grid. Thus GFO in power grids may be induced by wind power stochastic fluctuations.

5.2.3.3 Computational Analysis of System Oscillation Induced by Wind Power Fluctuation

(1) **System Modal Analysis**

This section adopts a power system simulation model conducted on the basis of the power system analysis software package (PSASP). This model contains 8587 buses, 883 generators and 623 stabilizers. It is used to simulate the large-scale AC–DC hybrid interconnected power grid, the North China-Central China Power Grid. Power flow interface data at 22:45:00 on January 1, 2015 are selected as an initial running status of the system when calculated.

A method of partial eigenvalues is performed to search for eigenvalues within the mentioned frequency range. The requisite oscillation modes are selected. Based on the general forced oscillation theory, only the system oscillation mode covered by the frequency range of wind power stochastic fluctuation, principally between 0.1 and 0.3 Hz, may be excited. Therefore, the search for system eigenvalues should be carried out within 0.1–0.3 Hz. It is indicated in the result that, in this way, there is an inter-area oscillation mode in North China-Central China at a frequency of 0.175 Hz, as shown in Fig. 5.24. It can be seen that masses of generators in Longgang, Duobaoshan, Qinbei, and Mengjin belonging to Henan province all participate in this oscillation mode and their participation is high.

(2) **Calculation Results and Contrastive Analysis of Measured Oscillation**

To ensure the great accuracy of calculation results, the computation time and the step size are set as 40 s and 0.01 s respectively. The wind power fluctuation data measured from 22:45:00 to 22:45:40 on January 1, 2015 are selected as the input disturbance. The disturbance injection point is the wind power fluctuation time-domain curve, as shown in Fig. 5.25a. The corresponding power spectrum is shown in Fig. 5.25b. It should be noted that it is not enough to consider the stochastic power fluctuation in a single wind farm for a large-scale interconnected power system. By the end of 2014, the total installed wind power capacity in the North China-Central China Power Grid has already surpassed 30,000 MW, while the maximum power generation is over 20,000 MW. However, no more than 2 power injection points are allowed in PSASP. Thus, in the computation procedure, the measured wind power data should be expanded by 380 times to represent the entire wind power fluctuation in the whole North China-Central China Power Grid.

Wind power is injected into a simulation system by a current injection method to analyze power spectral properties of the system response. The selected injection points of power fluctuation are at the Jiazhuang 220 kV bus and Shaohua 220 kV bus as these two wind farms are electrically close to Chang-Nan Line. To analyze the power system oscillation, the relative power angle curve of generators highly participating in the system oscillation mode and the power fluctuation curve of the inter-area tie line are commonly selected as output variables. So the active power in Chang-Nan Line is selected as an output variable, of which the time-domain curve and corresponding power spectrum are shown in Fig. 5.26. Meanwhile, the power

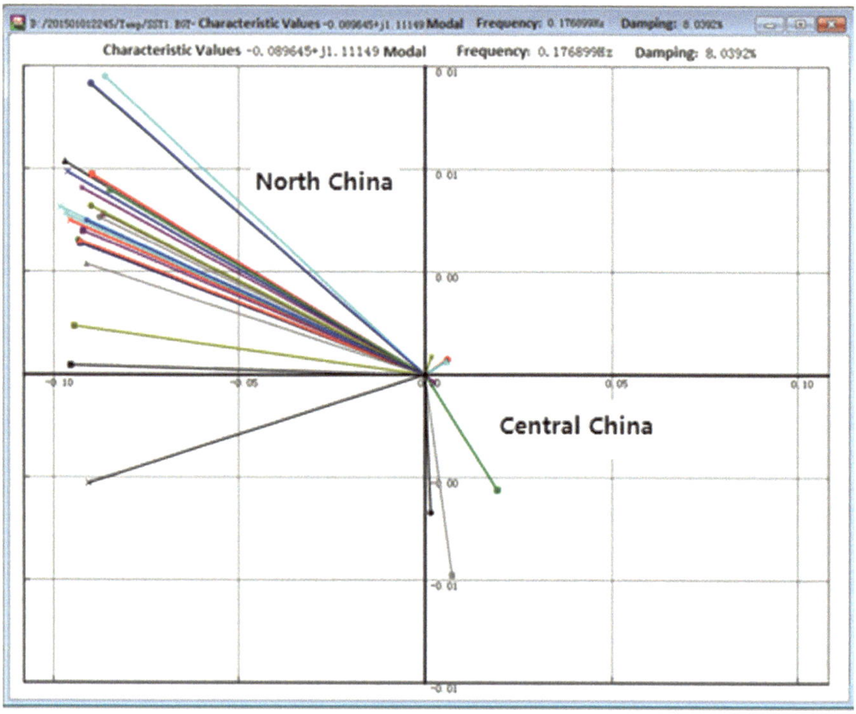

Fig. 5.24 Inter-area oscillation mode in North China-Central China and its major participant generators

fluctuation time-domain curve of Chang-Nan Line along with its power spectrum, measured from 22:45:00 to 22:45:40 on January 1, 2015, are given in Fig. 5.27 for comparison with the simulation results.

It can be concluded in Figs. 5.25, 5.26 and 5.27 that:

(1) The energy of wind power stochastic fluctuation is mainly distributed between 0.1 and 0.3 Hz which covers the system oscillation mode frequency 0.175 Hz. It can be seen that in the active power spectrum of the inter-area tie line, the system output energy is also distributed within 0.1–0.3 Hz, while an extremum occurs at 0.175 Hz. It means that there may be an internal link between the North China-Central China inter-area oscillations in the system and the stochastic fluctuation disturbance of wind power, which satisfies the conditions of GFO.
(2) Through the comparative analysis, fluctuations of the simulated curve differ from those of the measured curve in time-domain, but both frequency-domain properties are basically coincident. Both of their power spectra have similar peak properties and their peak values occurred at 0.175 Hz, which is equal to the frequency of the North China-Central China oscillation mode. The conclusion that can be drawn is that the stochastic power fluctuation frequently

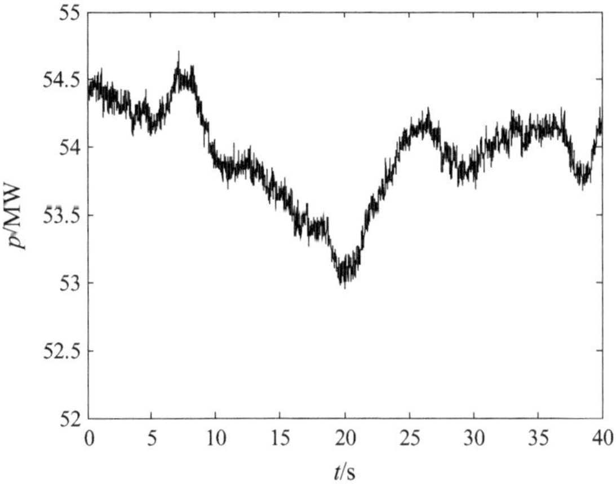

(a) Wind power time-domain fluctuation curve

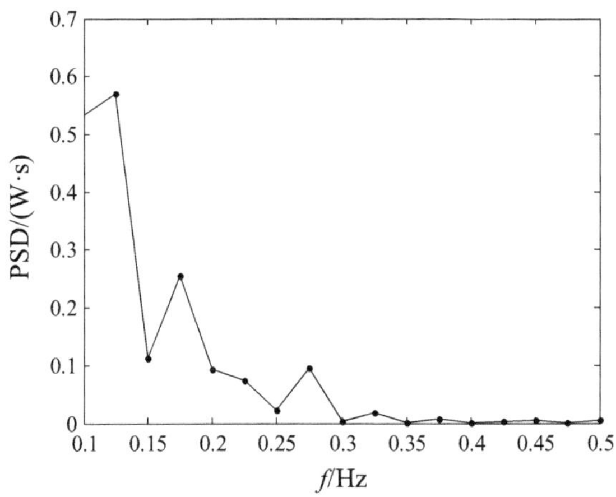

(b) Power spectrum of wind power frequency domain

Fig. 5.25 Wind power fluctuation time-domain curve and its power spectrum

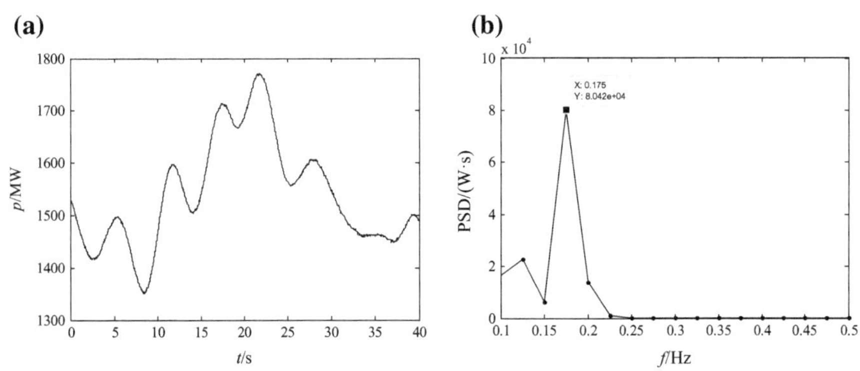

Fig. 5.26 Simulated time-domain curve of power fluctuation in Chang-Nan Line and its power spectrum

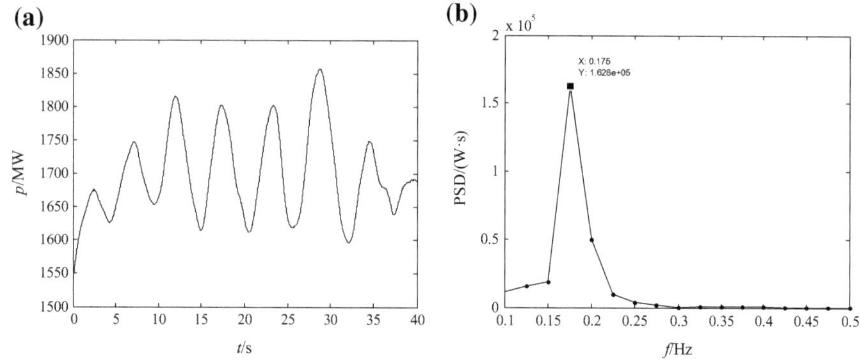

Fig. 5.27 Measured time-domain curve of power fluctuation in Chang-Nan Line and its power spectrum

occurring in Chang-Nan Line presumably is the general forced oscillation induced by stochastic disturbance such as wind power. The fluctuation can be explained by the general forced oscillation mechanism.

5.3 General Internal Resonance under Large Stochastic Disturbance

5.3.1 Interaction Mechanism Between Modes

The mode analysis and superposition principle are important method and principle to analyze the dynamic phenomenon of a multi-machine power system. Given certain assumptions, the decoupling between modalities by corresponding linear transformation makes the oscillation of the multi-machine system equivalent to the superposition of a series of modal oscillations. But this method ignores the influence of nonlinear high-order terms in the system state equation, and the occurrence of internal resonance just derives from the non-linear interaction between system oscillation modes. For a nonlinear multi-machine power system, the general form of its state equation is:

$$\frac{\mathrm{d}}{\mathrm{d}t}\boldsymbol{x}=\boldsymbol{F}(\boldsymbol{x}) \tag{5.3.1}$$

Wherein $\mathbf{x} \in \mathbf{R}^n$, $\boldsymbol{F}(\boldsymbol{x}) = [f_1(\boldsymbol{x}), f_2(\boldsymbol{x}), \ldots, f_n(\boldsymbol{x})]^T$ is a non-linear function. Performing a Taylor expansion at the equilibrium point $\boldsymbol{x}_{SEP}$ results in:

$$\frac{\mathrm{d}}{\mathrm{d}t}\boldsymbol{x} = \boldsymbol{A}\boldsymbol{x} + \frac{1}{2}\begin{bmatrix} \boldsymbol{x}^T\boldsymbol{H}^1\boldsymbol{x} \\ \vdots \\ \boldsymbol{x}^T\boldsymbol{H}^n\boldsymbol{x} \end{bmatrix} + O(\|\boldsymbol{x}\|^3) \tag{5.3.2}$$

Wherein the first term on the right side of the equation is a linear term, and $\boldsymbol{A}$ represents the Jacobian matrix of $\boldsymbol{F}'(\boldsymbol{x})$ at the equilibrium point $\boldsymbol{x}_{SEP}$. The second term is a second-order nonlinear term, and $\boldsymbol{H}^i$ represents the Hessian matrix of $f_i(\boldsymbol{x})$ at the equilibrium point $\boldsymbol{x}_{SEP}$. The third term is a high-order nonlinear term of third or higher order.

In the system oscillation analysis and study, modal coordinates are generally used. A new coordinate equation can be formed by substituting the given terms $\boldsymbol{x} = \boldsymbol{\Phi} z$, $\boldsymbol{\Psi} = \boldsymbol{\Phi}^{-1}$ into Eq. (5.3.2):

$$\frac{\mathrm{d}}{\mathrm{d}t}z = \Lambda z + \frac{1}{2}\boldsymbol{\Psi}^T\begin{bmatrix} z^T\boldsymbol{\Phi}^T\boldsymbol{H}^1\boldsymbol{\Phi}z \\ \vdots \\ z^T\boldsymbol{\Phi}^T\boldsymbol{H}^n\boldsymbol{\Phi}z \end{bmatrix} + O(\|z\|^3) \tag{5.3.3}$$

Wherein $\boldsymbol{\Psi}, \boldsymbol{\Phi}$ are left and right eigenvector matrices of $\boldsymbol{A}$, respectively. Λ are diagonal matrices composed of the eigenvalues λ_i. From the above equation, due to the nonlinear terms, various modes of the system cannot be completely decoupled, which may cause more complex nonlinear phenomena. According to the Eq. (5.3.3), for an oscillation mode m in a nonlinear power system, the expression is:

$$\frac{\mathrm{d}}{\mathrm{d}t}z_m = \lambda_m z_m + \sum_{i=1}^{n}\sum_{j=1}^{n} c_{ij}^m z_i z_j + O(\|z\|^3) \tag{5.3.4}$$

Wherein, c_{ij}^m represents the corresponding element in the Hessian matrix $\boldsymbol{H}^m$.

The linear analysis method mainly analyzes the first term on the right side of Eq. (5.3.4), while the interaction between modes is mostly reflected on the second term and the third term on the same side. Since the nonlinearity of the system usually decreases from low order to high order, the current research mainly focuses on the second-order nonlinearity term in the Eq. (5.3.4). Similarly, to the analytical method of forced oscillation, the nonlinear term can be considered as a "disturbance" to the mode m after composite transformation of the system various oscillation modes. It will affect the system oscillation if certain conditions are met. The expression is as follows:

$$g_m(z_1, z_2, \ldots, z_n) = \sum_{i=1}^{n}\sum_{j=1}^{n} c_{ij}^m z_i z_j \tag{5.3.5}$$

It shows that the disturbance of the second-order nonlinear term for the oscillation mode m mainly comes from the oscillation behaviors of the other modes of the system. Researches have showed that for a free-oscillation condition after system disturbance, when the system is excited to cause multiple modes of oscillations and there is an appropriate coupling between two oscillation modes, oscillations of the two modes might become gains of each other. Even the oscillations that should have rapidly attenuated under the damping might attenuate first and then rapidly increase.

The power system oscillation under the stochastic disturbance is an active forced oscillation. Under the continuous function of a disturbance source, the oscillation does not decay rapidly under the damping effect. At this time, the excited oscillation mode will have more obvious effects on other modes of the system. Assuming that one of the system oscillation modes k is excited and other modes have less effects on the mode m, only the effect of mode k on the mode m is taken into account in Eq. (5.3.5), which can be indicated by $g_m(z_k)$. This chapter centers on the following two cases of $g_m(z_k)$:

(1) In the Eq. (5.3.5) $z_i = z_j = z_k$,

$$g_m(z_k) = c_{kk}^m z_k^2 \tag{5.3.6}$$

(2) In the Eq. (5.3.5) $z_i = z_k,\ z_j = z_{\bar{m}}$,

$$g_m(z_k) = c_{k\bar{m}}^m z_k z_{\bar{m}} \tag{5.3.7}$$

Wherein the mode $\bar{m}$ indicates a conjugate mode of the mode m. If the mode k is excited to oscillate, it is equivalent to applying a disturbance $g(z_k)$ to mode m under the effect of a nonlinear term. At this point, if certain conditions are met, the mode m will be excited, and a significant oscillation is generated.

In case of a free oscillation or special forced oscillation, when $g(z_k)$ in the Eq. (5.3.6) is considered, if oscillation mode k is excited to generate a periodic forced oscillation, and the ratio of inherent frequencies of the oscillation mode k and mode m is 1:2, the internal resonance phenomenon will occur which is called the 1:2 internal resonance [9]. More generally, if three or more nonlinear terms are considered, then 1:1, 1:3 or more complex internal resonance phenomena exist. Under a stochastic disturbance, the 1:2 internal resonance phenomenon of the power system and its starting conditions will change due to the stochastic excitation of the external disturbance and the narrow-band power spectrum. Therefore, this chapter refers to internal resonance under stochastic disturbance as "generalized internal resonance" instead of following the former name of "1:2 internal resonance". In addition, according to the different cases represented by Eqs. (5.3.6) and (5.3.7), the generalized internal resonance can be divided into two types: "double-frequency internal resonance" and "half-frequency internal resonance". They are to be analyzed and verified respectively.

5.3.2 Generalized Double-Frequency Internal Resonance under Large Stochastic Disturbance

5.3.2.1 Mechanism Analysis

Suppose that under certain narrow-band stochastic disturbances, the system oscillation mode k is excited and the other mode not because its frequencies are not covered. The mode k output z_k is still a narrow-band stochastic process. Although the width of the z_k power spectrum is consistent with the width of the external disturbance spectrum, the power spectral characteristics of z_k under the action of the frequency-domain transfer function of the system have undergone some changes. It can be observed that most of the oscillation energy of z_k is mainly centered in the small range Δf around the frequency f_k of the mode k, and the extreme value appears at f_k, while the energy in the rest of the frequency range is smaller. In order to choose a simple example for analysis, it is assumed that the z_k power spectrum $S_k(f)$ is shown as Fig. 5.28a, and therefore the expression is as follows:

$$S_k(f) = \begin{cases} c, & f_k - \frac{\Delta f}{2} < f < f_k + \frac{\Delta f}{2} \\ 0, & Others \end{cases} \tag{5.3.8}$$

Wherein: c is a constant; f_k is a natural oscillation frequency of the oscillation mode k.

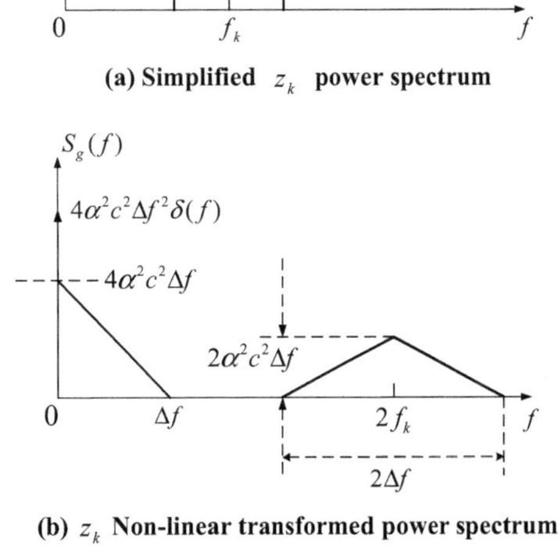

Fig. 5.28 Narrow-band stochastic process power spectrum

Because the external stochastic disturbance has been there, z_k cannot be suppressed. When considering the nonlinear coupling between mode k and mode m, it is equivalent to transforming the stochastic process z_k nonlinearity into $g_m(z_k)$, which $g_m(z_k)$ can be regarded as a stochastic disturbance continuously applied on mode m. If only the case of (5.3.6) is considered, and given that $\alpha = c_{kk}^m$, then $g_m(z_k) = \alpha z_k^2$, and $g_m(z_k)$ power spectrum $S_g(f)$ obtained by analytical analysis is:

$$S_g(f) = \begin{cases} 4\alpha^2 c^2 \Delta f^2 \delta(f), & f = 0 \\ 4\alpha^2 c^2 (\Delta f - f), & 0 < f < \Delta f \\ 2\alpha^2 c^2 (\Delta f - |f - 2f_k|), & 2f_k - \Delta f < f < 2f_k + \Delta f \\ 0, & Others \end{cases} \tag{5.3.9}$$

$g(z_k)$ power spectrum $S_g(f)$ is as shown in Fig. 5.28b.

It can be seen that the energy of $g(z_k)$ is concentrated in the two frequency bands $[0,\ \Delta f)$ and $(2f_k - \Delta f,\ 2f_k + \Delta f)$, which means that when the mode k is excited to generate a continuous forced oscillation z_k, under the effect of the square nonlinear coupling between system modes, it is equivalent to applying a stochastic disturbance $g(z_k)$ with a power spectral range $[0,\ \Delta f) \cup (2f_k - \Delta f,\ 2f_k + \Delta f)$ to the system. According to the general forced oscillation theory, if the frequency of an oscillation mode m satisfies the condition shown in Eq. (5.3.10), the mode m will be excited to produce an internal resonance phenomenon.

$$f_m \in [0,\ \Delta f) \cup (2f_k - \Delta f,\ 2f_k + \Delta f) \tag{5.3.10}$$

It should be noted that, as Δf is small, $f_m \in [0,\ \Delta f)$ is less likely to happen, so only the situation $f_m \in (2f_k - \Delta f,\ 2f_k + \Delta f)$ is worth focusing on. In this case, since $g(z_k)$ power spectrum amplitude has an extreme value at $2f_k$, when $f_m = 2f_k$, the amplitude of the caused oscillation reaches its peak. Although this is the result of a simple example, it is in fact generic.

From this, it can be seen that under a narrow-band stochastic disturbance, the phenomenon of internal resonance may occur without having frequencies between system oscillation modes reaching strict integral multiples. This conclusion extends the initiation condition of internal resonance under the free-oscillation or special-forced oscillation. That is, the internal resonance phenomenon of the power system is more easily excited under the stochastic disturbance. Therefore, this type of internal resonance phenomenon is called “double frequency internal resonance”.

5.3.2.2 Example Verification

The example system used in this section is a 3-generator 9-bus system on the Western United States Grid (WSCC), as shown in Fig. 5.29. As the inertia time constant of generator G1 is larger, it is set as a reference machine. Only two electromechanical oscillation modes in the system are conducive to maximize the coupling between the modes by parameter adjustment and so on, so as to affect the oscillation response of the system more directly. At the same time, the system response interference of the coupling between other modes can be avoided.

In the example system, the generator adopts a fifth order detailed model and the load adopts a constant impedance model. Using a small signal analysis module in PSASP, the system's 2 electromechanical oscillation modes and the main participating units will be obtained. The inertia time constant of the generator is properly adjusted, and the load level increases along with the active power output. The system oscillation mode to the condition of weak damping and frequency ratio close to 1:2 is set as shown in Table 5.6. At this moment, the rated active output of generators G1, G2 and G3 is 570, 300 and 200 MW, respectively. The degree to which each unit participates in each oscillation mode is decreasingly arranged from left to right at both ends of the "↔" symbol.

It should be noted that the natural oscillation frequencies of the example system mode 1 and mode 2 do not satisfy exactly 1:2, but that there is a deviation of about 0.1 Hz. Firstly, considering the system excited by sinusoidal disturbance with frequency 0.54 Hz and amplitude 50 MW injected at the outlet of generator G3, the relative power angle response power spectrum between generator G2 and reference generator G1 is shown in Fig. 5.30. It can be seen that under the strong sinusoidal disturbance, Mode 1 is excited to give a large oscillation because the disturbance frequency is equal to the frequency of Mode 1. Under the secondary non-linear coupling between these modes, it is equivalent to generating a sinusoidal disturbance source with a frequency of 1.08 Hz. However, since the disturbance source frequency deviates from the frequency of Mode 2, Mode 2 is not excited, and only a small oscillation with a frequency of 1.08 Hz is generated. Therefore, because the system mode frequency does not satisfy the 1:2 relationship, the internal resonance phenomenon will not arise in the system under strong sinusoidal disturbance.

Secondly, in order to verify the phenomenon of double frequency internal resonance described in the previous section, the case where the system is subject to a narrow-band stochastic disturbance is considered. It is assumed that the power spectrum band of the externally applied stochastic disturbance lies between 0.4 and 0.6 Hz and the remaining frequency bands are 0. This means that only the natural

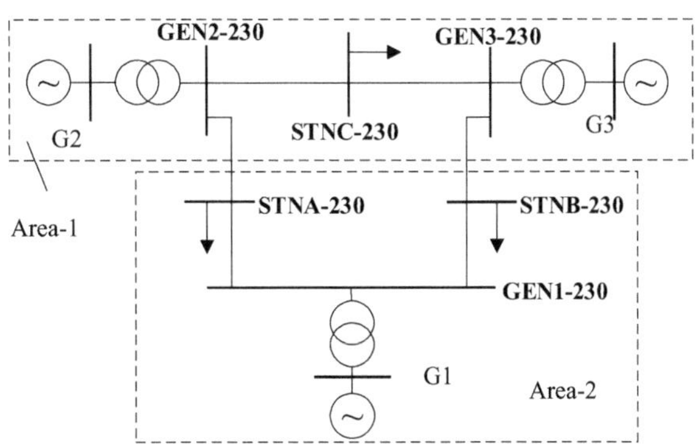

Fig. 5.29 Example system diagram

Table 5.6 System oscillation mode and main participating units

Mode	Frequency/Hz	Damping ratio/%	Participating unit
1	0.5418838	4.4007	G_3, $G_2 \leftrightarrow G_1$
2	0.9705482	4.0304	$G_2 \leftrightarrow G_3$

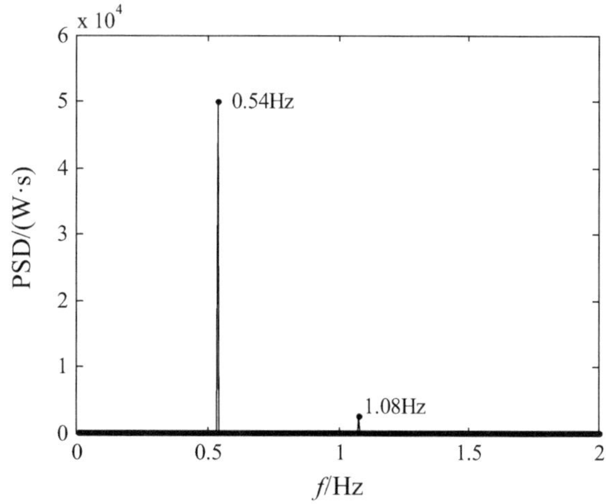

Fig. 5.30 G2 relative power angle power spectrum

oscillation frequency of the oscillation mode 1 is covered. Then the frequency-time transform method is used to obtain a time-domain sequence of stochastic disturbance, and the stochastic disturbance is injected into the generator G3 outlet bus by current injection to simulate the stochastic fluctuations of the generator's active power output. The relative power angles of generators G2 and G3 and reference generator G1, are observed respectively, and their power spectrums are analyzed.

Figure 5.31 shows the power spectrum of the stochastic disturbance and the obtained time-domain sequence. The active power p in the figure is a per unit value (1 p.u. = 100 MW) (the same as below). From Fig. 5.31b, it can be seen that the fluctuation range of stochastic disturbance lies between −1.5 and 1.5 MW, and its disturbance intensity is weaker than that of a rated active output of each generator.

On the aspect of time-domain fluctuations in the system response, the generator relative power angle fluctuation range is also small, G2's fluctuation ranges around 0.7°, and G3's fluctuation ranges around 2.5°. In addition, oscillations of the two generators are almost identical, as shown in Fig. 5.32. The power spectrum of the power angle response curves of the generators G2 and G3 are analyzed respectively. The analysis results are shown in Fig. 5.33.

Figure 5.33 shows that under stochastic disturbance, the system oscillation mode 1 is excited, and the power angle response power spectrum of the generator shows an extreme value at 0.54 Hz. Because both generators G2 and G3 are involved in Mode 1, the oscillations of both can be consistent, but the two oscillations have different intensities due to the degree of involvement. At the same time, it can be

Fig. 5.31 Stochastic disturbance power spectrum and time domain sequence

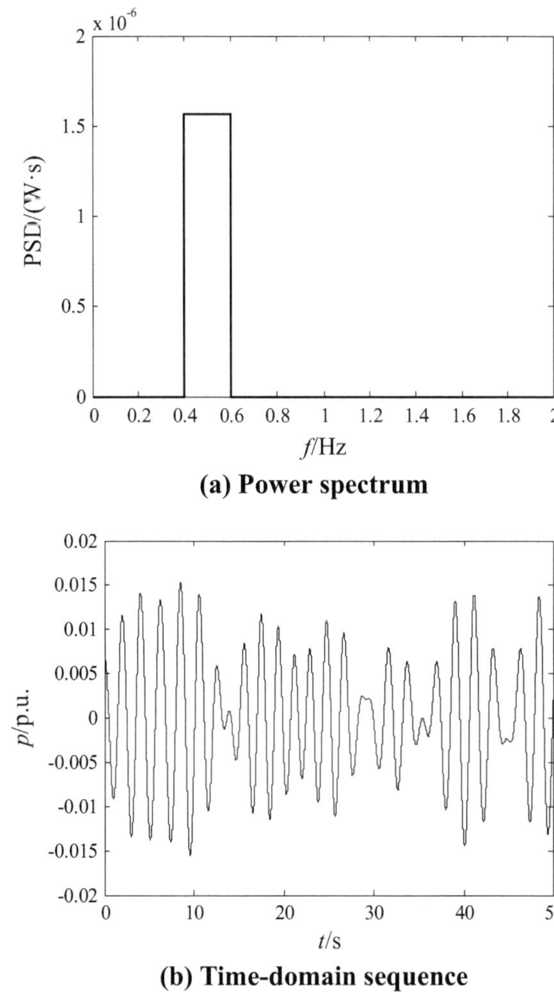

found that the phenomenon of double frequency internal resonance does not appear in the system because the stochastic disturbance intensity is weak and the system does not show strong nonlinear characteristics. Therefore, the stochastic disturbance intensity is considered to be increased to stimulate the double frequency internal resonance in the system.

In order to facilitate comparison, apart from increasing the stochastic disturbance intensity, the same set of stochastic sequences is employed. In this way, as the active power output fluctuation range of the generator is increased, the consistency of the power variation in the two simulations can be guaranteed. Figure 5.34 shows the power spectrum of the stochastic disturbance and the obtained time-domain sequence. At this point, the fluctuation range of the stochastic disturbance lies between −50 and 50 MW.

Fig. 5.32 Generator relative power angle curve under stochastic disturbance

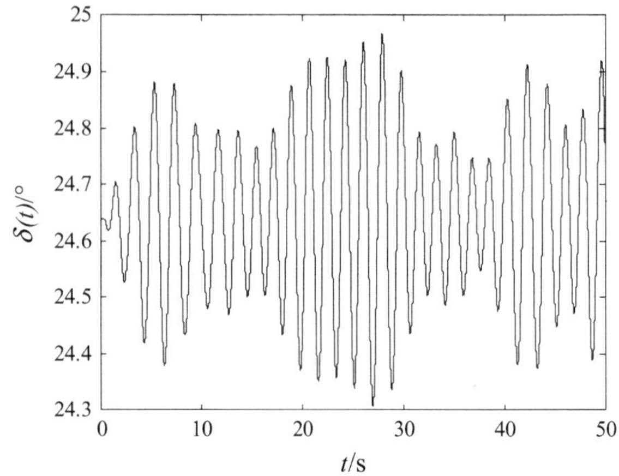

(a) G2 relative power angle

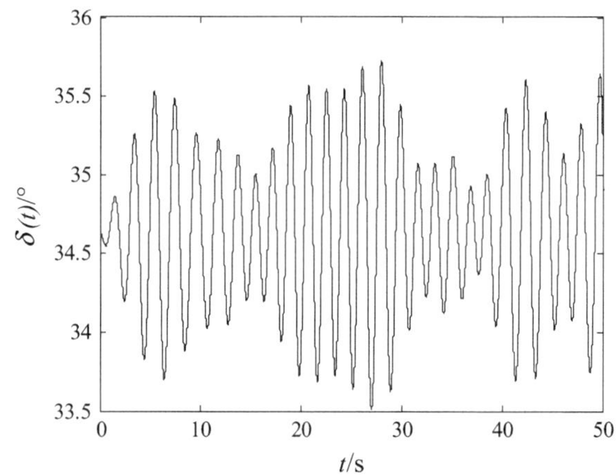

(b) G3 relative power angle

Figure 5.35 shows the generator relative power angle response curve. Comparing the power angle responses of generators G2 and G3, it is found that both power angle oscillation ranges show different degrees of increase, which is attributed to the increase of disturbance intensity. At the same time, compared with the simulation results in Fig. 5.32, it is demonstrated that the two oscillations of generator G3 basically share a same regularity. Additionally, the oscillation regularity of generator G2 changes and the power angle oscillation curve shows obvious high-frequency component.

Similarly, the power spectrums of the power angle response curves shown in Fig. 5.35 are analyzed. The results of this analysis are shown in Fig. 5.36. It is revealed that there are three frequency band components in the response power spectrum of generator G2, which are around 0 Hz, around 0.54 Hz and around

Fig. 5.33 Generator relative power angle power spectrum

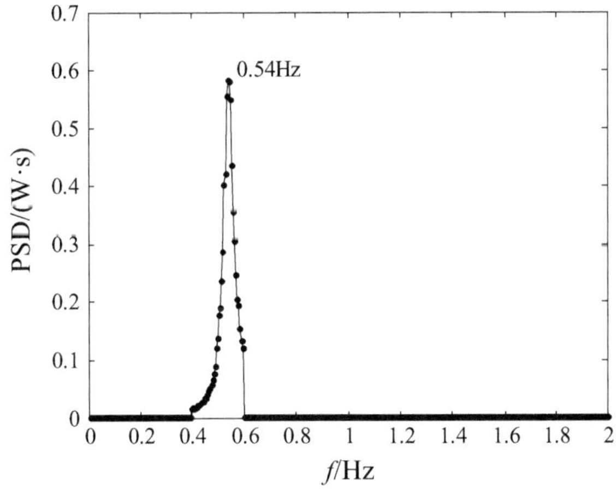

(a) G2 relative power angle power spectrum

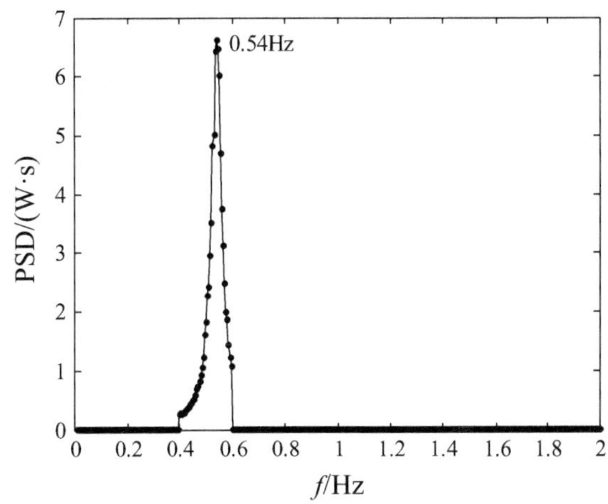

(b) G3 relative power angle power spectrum

0.97 Hz respectively, as shown in Fig. 5.36a. In the response power spectrum of generator G3, the component of the frequency band around 0.54 Hz can be clearly observed, as shown in Fig. 5.36b. By partially amplifying the response power spectrum of the generator G3, the frequency band components around 0 and 0.97 Hz are also observed but the amplitudes are relatively small, as shown in Fig. 5.36c. Therefore, from the result of power spectrum analysis, the power angle oscillation curves of generators G2 and G3 both show the components of three frequency bands and the extreme values at 0.54 and 0.97 Hz. They indicate that Mode 1 and Mode 2 are excited to show a double frequency resonance phenomenon.

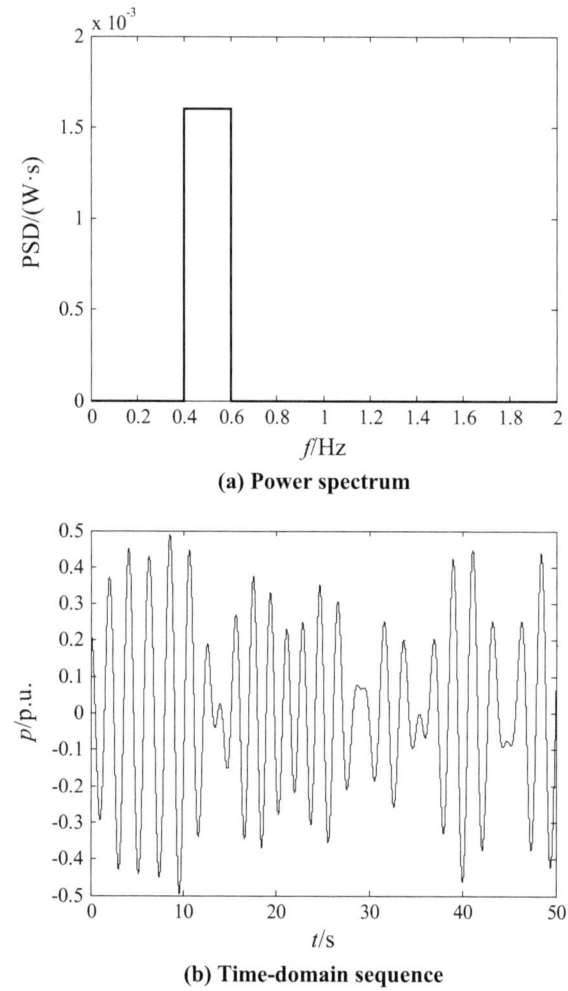

Fig. 5.34 Stochastic disturbance power spectrum and time domain sequence

From this it is concluded that under strong stochastic disturbance, the system shows a strong nonlinear characteristic, and thus the phenomenon of double frequency internal resonance has occurred. The Mode 2 is excited to show an obvious oscillation. At this time, the system response energy is distributed over a small range near zero frequency, Mode 1 frequency and Mode 2 frequency under the effect of non-linear coupling between modes. This phenomenon validates the theoretical analysis results of the last section.

At the same time, it is also seen that when the disturbance intensity is increased to a certain extent, the frequency component of Mode 2 in G2's power angle response power spectrum will be larger than that of the Mode 1. This is attributable to the increase of Mode 2 oscillation energy caused by the enhancement of the nonlinearity of the system, and the higher involvement of generator G2 in Mode 2.

Fig. 5.35 Generator relative power curve under stochastic disturbance

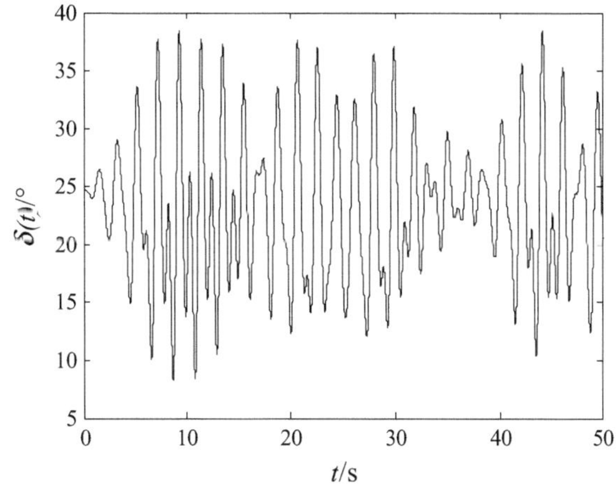

(a) G2 relative power angle

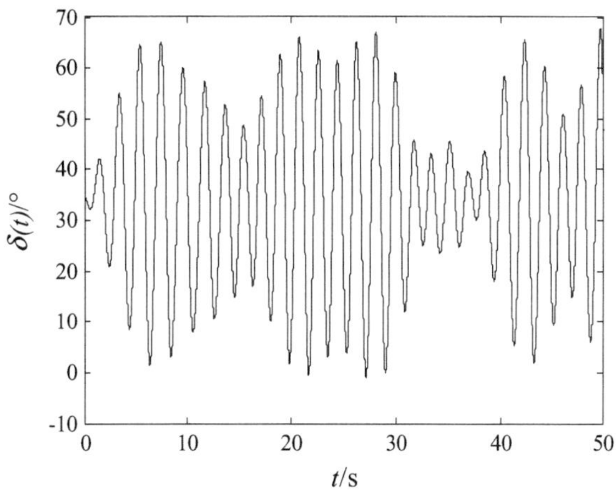

(b) G3 relative power angle

This means that if a similar situation happens to a generator power angle in a real power grid and the oscillation at Mode 2 is more pronounced, the oscillation at Mode 1 may be easily neglected when taking control measures for suppression. Otherwise, unnecessary losses would be caused when only investigating possible disturbance sources for Mode 2 frequencies.

Comparing the initiation conditions of double frequency internal resonance under different situations, it can be seen that under the special forced oscillation, if the system oscillation Mode 1 and Mode 2 cannot meet $f_2 = 2f_1$, the phenomenon of double frequency internal resonance cannot appear. Under the narrow-band stochastic disturbance, when $f_2 \in (2f_1 - \Delta f,\ 2f_1 + \Delta f)$ is satisfied, Mode 2 can be excited, and then double frequency internal resonance appears.

Fig. 5.36 Generator relative power angle power spectrum

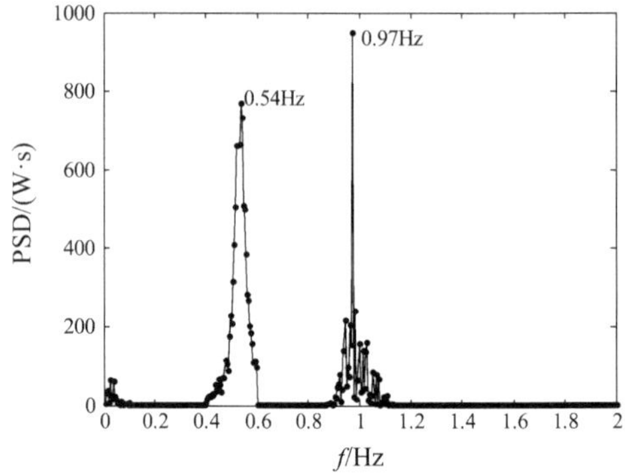

(a) G2 relative power angle power spectrum

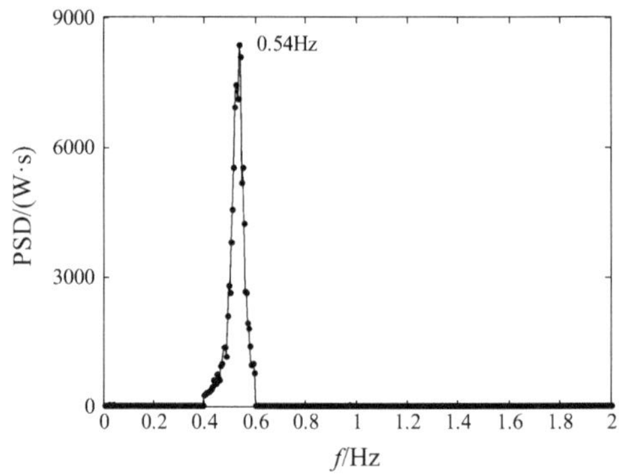

(b) G3 relative power angle power spectrum

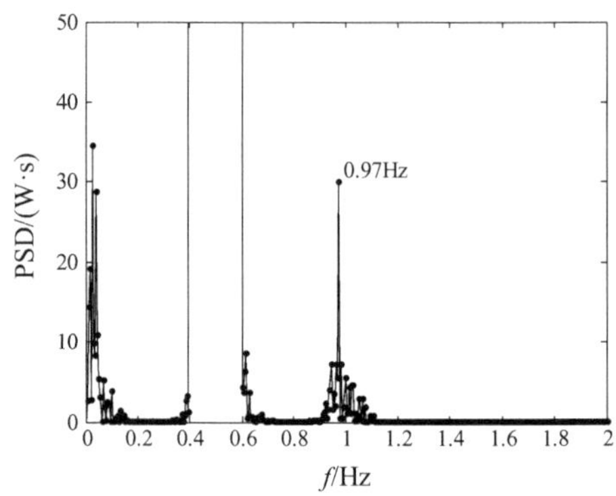

(c) G3 relative power angle power spectrum (local amplification)

5.3.3 General Half-Frequency Internal Resonance under Large Stochastic Disturbance

5.3.3.1 Mechanism Analysis

The mechanism of double frequency internal resonance reveals the possibility of high frequency mode oscillations induced by low frequency mode oscillation phenomena in the power system. According to the definition and classification of oscillations, compared to the higher frequency, the lower frequency inter-area oscillation mode is more destructive to a power system due to its wide range of oscillation and higher participation in the generators. When the high-frequency local mode oscillation is firstly excited, if the low-frequency inter-area mode oscillation can be induced by the nonlinear mode coupling, then this situation may have a greater and more significant impact on the system than that of the double frequency internal resonance. This is worth our attention. For this reason, in this chapter, this type of internal resonance phenomenon is called half-frequency internal resonance, and tentative analyses and explanations of this oscillation mechanism are carried out.

The mechanism explanation for the phenomenon needs to take into account the coupling between modes as shown in Eq. (5.3.7). In contrast to the coupling for double-frequency internal resonance, Eq. (5.3.7) focuses on the effect of the interaction of mode k oscillation and conjugate mode $\bar{m}$ oscillation products excited by a disturbance source on mode m. When the power system is disturbed, all the oscillation modes of the system may be excited to different degrees during the oscillation start-up phase in the system. The mode subject to a continuous disturbance source will manifest a long time forced oscillation and later become the dominant oscillation mode. While oscillation amplitudes of the other modes will be gradually reduced to 0, the oscillation decay time is relevant to the damping of the mode. It should be noted that since the eigenvalues of the system state equation are paired in conjugate form, and each pair corresponds to one oscillation mode, the "mode k oscillation" mentioned in this chapter is actually superposed of modes k and their conjugate mode $\bar{k}$ responses. The power spectrum in the frequency domain has a centrosymmetric characteristic with respect to the frequency point of 0 Hz, and so does the "Mode m oscillation". When the conjugate mode is not involved, usually only the positive frequency of practical physical meaning is considered, i.e. the unilateral spectrum form of the power spectrum is analyzed.

Assuming that under a certain narrow-band stochastic disturbance the oscillation mode k is excited and the output of the time domain is a narrow-band stochastic process, the power spectrum analysis of this scenario generally focuses only on the power spectrum of the mode response z_k, ignoring conjugate mode response $z_{\bar{k}}$ power spectrum in the negative frequency range, as shown in Fig. 5.37a below. At the same time, a weakly damping oscillation mode m is excited to show a short period of periodic decay oscillation in the start-up phase. Its time domain output can be expressed as a superposition of the mode m response and the conjugate mode $\bar{m}$ response, and the expression of both are as follows:

$$\begin{cases} z_m = z_{m0} \cdot e^{\lambda_m t} = z_{m0} \cdot e^{-\sigma_m t} \cdot e^{j2\pi f_m t} \\ z_{\bar{m}} = z_{\bar{m}0} \cdot e^{\lambda_{\bar{m}} t} = z_{\bar{m}0} \cdot e^{-\sigma_{\bar{m}} t} \cdot e^{j2\pi f_{\bar{m}} t} \end{cases} \tag{5.3.11}$$

Wherein z_{m0}, $z_{\bar{m}0}$ show the initial values of the response of the mode m and the conjugate mode $\bar{m}$. λ_m, $\lambda_{\bar{m}}$ represent a pair of conjugate eigenvalues of the oscillation mode m. The response power spectrum of the conjugate mode $\bar{m}$ is shown in Fig. 5.37b.

Substituting Eq. (5.3.11) into Eq. (5.3.7) results in:

$$g_m(z_k) = c^m_{k\bar{m}} z_k z_{\bar{m}} = c^m_{k\bar{m}} z_{\bar{m}0} e^{-\sigma_{\bar{m}} t} \cdot z_k e^{j2\pi f_{\bar{m}} t} \tag{5.3.12}$$

With reference to the analysis method of the double frequency resonance mechanism, the Eq. (5.3.12) is analyzed in respect of the frequency domain. The multiplication of time-domain signals can be transformed into convolution in the frequency domain. According to the convolution theorem and the frequency shift theorem, if a signal is multiplied by a complex factor $e^{j2\pi f_0 t}$ in the time domain, it is equivalent to the distance of the spectrum of the signal shift f_0 in the frequency domain. Therefore Eq. (5.3.12) is equivalent to frequency modulation on the basis of mode k oscillation z_k. The power spectrum features of the coupling between modes $g_m(z_k)$ are shown as Fig. 5.37c.

It can be seen that the $g_m(z_k)$ power spectrum does not change significantly and is still essentially a narrow-band stochastic process, except that the frequency range of its energy distribution has migrated as a whole, mainly in a small neighborhood domain centered at frequency $f_k - f_m$. According to the general forced oscillation theory, if the $g_m(z_k)$ power spectrum frequency band can cover the oscillation frequency of the mode m at this time, i.e. it meets $f_m \in \left[f_k - f_m - \frac{1}{2}\Delta f,\ f_k - f_m + \frac{1}{2}\Delta f\right]$, the mode m will be excited to generate the internal resonance phenomenon. After simplification, the conditions required to initiate the half-frequency internal resonance phenomenon are:

$$f_m \in \left[\frac{1}{2} f_k - \frac{1}{4}\Delta f,\ \frac{1}{2} f_k + \frac{1}{4}\Delta f\right] \tag{5.3.13}$$

The above analysis shows that low-frequency mode oscillations induced by high-frequency mode oscillations are possible. Under the narrow-band stochastic disturbance, the oscillation frequencies between each other in the system do not need to follow the strict integral multiple to initiate the phenomenon of half-frequency internal resonance. However, it should be pointed out that, compared with the conditions required for the system frequency oscillation mode in the double-frequency internal resonance, the half-frequency internal resonance has harsher starting conditions and relatively smaller initiation probability.

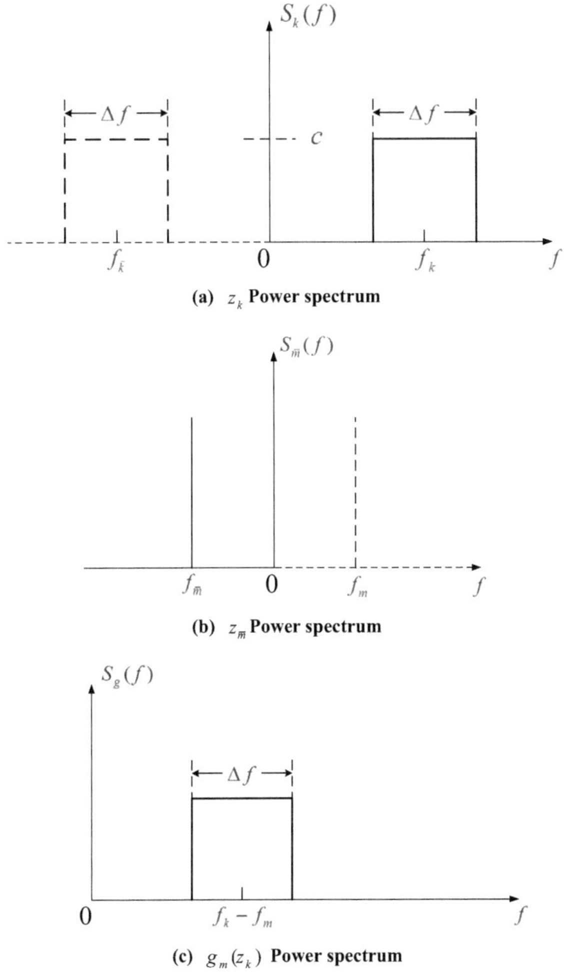

Fig. 5.37 Power spectral characteristics of coupling between modes

5.3.3.2 Example Verification

In this section the WSCC 3-generator and 9-bus system is used again for the calculation shown in Fig. 5.29. The model used for generator and load remains the same as well. Compared to the double frequency internal resonance, the half-frequency internal resonance has harsher starting conditions and relatively smaller initiation probability, thus the model parameters need to be further adjusted to weaken the system damping. Using the small signal analysis module in PSASP, the system's 2 electromechanical oscillation modes and the main participating units will be obtained as shown in Table 5.7.

At this moment, the rated active power output of generators G1, G2 and G3 is 570, 300 and 200 MW, respectively. The degree to which each unit participates in

Table 5.7 System oscillation mode and main participating units

Mode	Frequency/Hz	Damping ratio/%	Participating unit
1	0.6032557	1.3198	G_3, $G_2 \leftrightarrow G_1$
2	1.3117390	1.2550	$G_2 \leftrightarrow G_3$

each oscillation mode is decreasingly shown from left to right at both ends of the "↔" symbol. As shown in the table, the oscillation frequencies in the system do not meet the condition of the strict integral multiple.

Firstly, given the system excited by sinusoidal disturbance, a sinusoidal disturbance of 1.31 Hz frequency and 70 MW amplitude is injected at a load bus STNC-230. A relative power angle response power spectrum between generator G3 and reference generator G1 is shown in Fig. 5.38. At this time, the power spectrum energy is concentrated at the frequency of 1.31 Hz, and the energy at the remaining frequency points is zero. It is illustrated that under the strong sinusoidal disturbance, Mode 2 is excited and shows special forced oscillation because the disturbance frequency is equal to the frequency of mode 2. However, oscillation Mode 1 is not excited, and no half-frequency internal resonance occurs. That is because the system's oscillation mode frequencies fail to follow the strict integer multiple relation.

Secondly, in order to verify the phenomenon of half-frequency internal resonance described in the previous section, the case where the system is subject to a narrow-band stochastic disturbance is considered. It is assumed that the power spectrum band of the externally applied stochastic disturbance lies between 1.2 and 1.4 Hz and the amplitudes of the remaining frequency bands are 0. This means that only the natural oscillation frequency of oscillation Mode 2 is covered. Then the stochastic disturbance is injected into the STNC-230 bus to simulate the stochastic power fluctuation of the load with a power steady-state value of 600 MW. Finally, the power spectrum analysis of the relative power angles of the generators G2 and

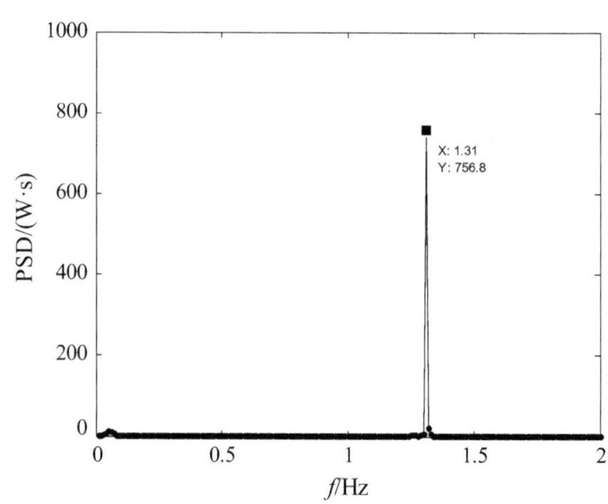

Fig. 5.38 Generator G3 relative power angle power spectrum

G3 with the reference generator G1 is carried out. Figure 5.39 shows the power spectrum of the stochastic disturbance and the obtained time-domain sequence. The active power p in the figure is a per unit value (1 p.u. = 100 MW), hereinafter the same.

Figure 5.39b shows that the fluctuation range of stochastic disturbance lies between −10 and 10 MW and that its disturbance intensity is weaker than that of the load power steady-state value. Figure 5.40 shows the relative power angle response curves of generators G2 and G3 respectively and the corresponding power spectra. It is observed that G2 and G3 are similar in the fluctuations of the angle response time-domain curves. In terms of the power spectrum, only the general

Fig. 5.39 Stochastic disturbance power spectrum and time domain sequence

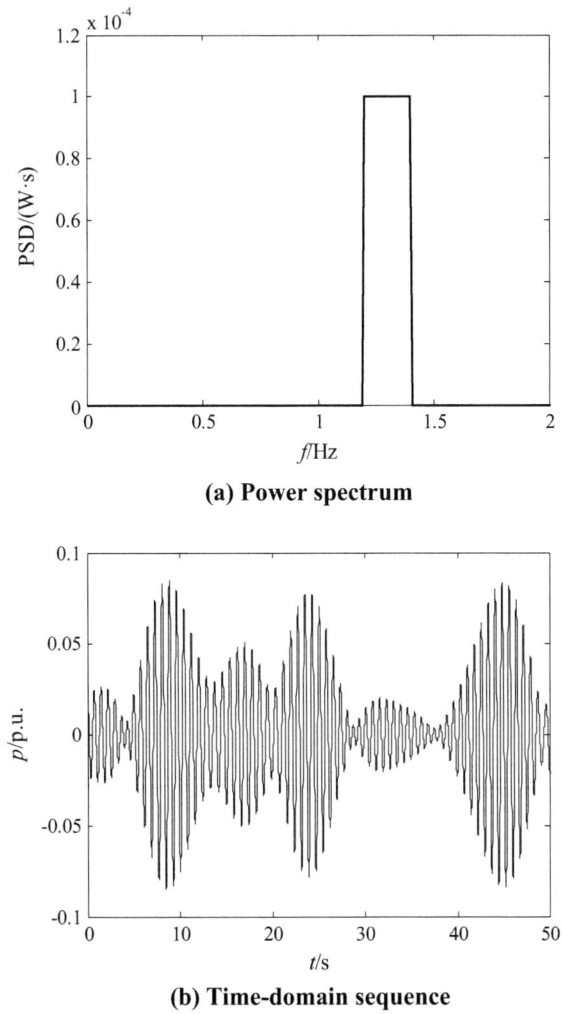

forced oscillation of Mode 2 is excited, while Mode 1 is not. Therefore, at this point, no half-frequency internal resonance phenomenon occurs.

In order to increase the intensity of stochastic disturbance to make the system show a stronger non-linear characteristic and to facilitate a comparative analysis, the same set of stochastic sequences is employed. The stochastic disturbance fluctuation range lies between −60 and 60 MW after the enhancement, and its power spectrum and the time-domain fluctuation curve are shown in Fig. 5.41. With the enhanced stochastic disturbance injected into the example system, the power angle response and power spectral characteristics of generators G2 and G3 are obtained as shown in Fig. 5.42.

As the intensity of stochastic disturbance increases, the power angle oscillation amplitude of the system goes up, which represents a stronger nonlinear characteristic. At this time, the power angle response fluctuations of generators G2 and G3 have changed to varying degrees. It can be seen especially from the G3 power angle time domain fluctuation curve that multiple frequency components are observed clearly therein. From the power spectrum of relative power angle, the Mode 1 frequency component of around 0.61 Hz appears in G3, which indicates that the nonlinear coupling between modes causes a half frequency internal resonance

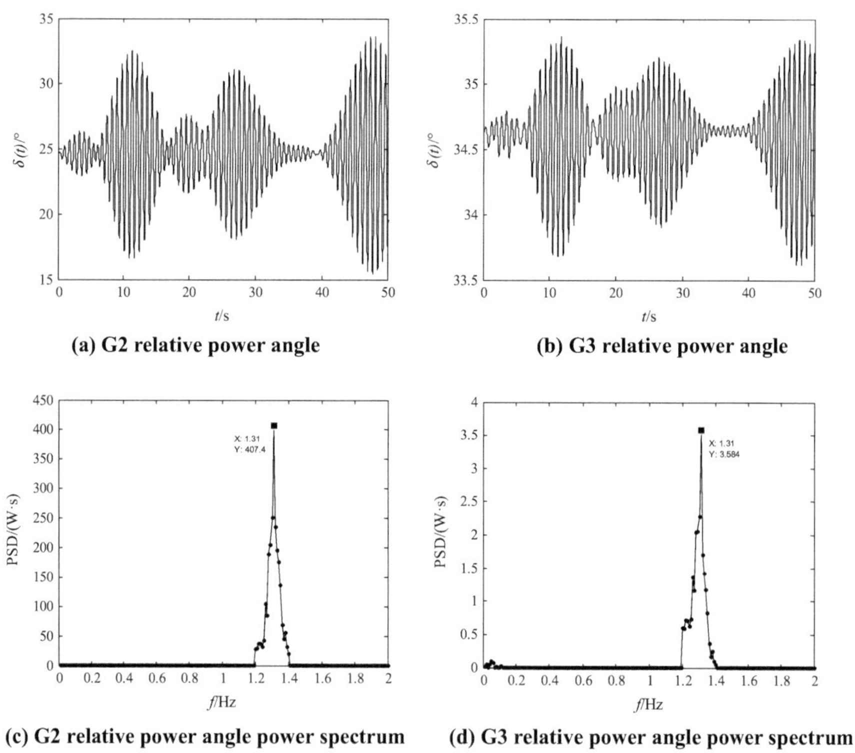

(a) G2 relative power angle (b) G3 relative power angle

(c) G2 relative power angle power spectrum (d) G3 relative power angle power spectrum

Fig. 5.40 Generator relative power angle power spectrum

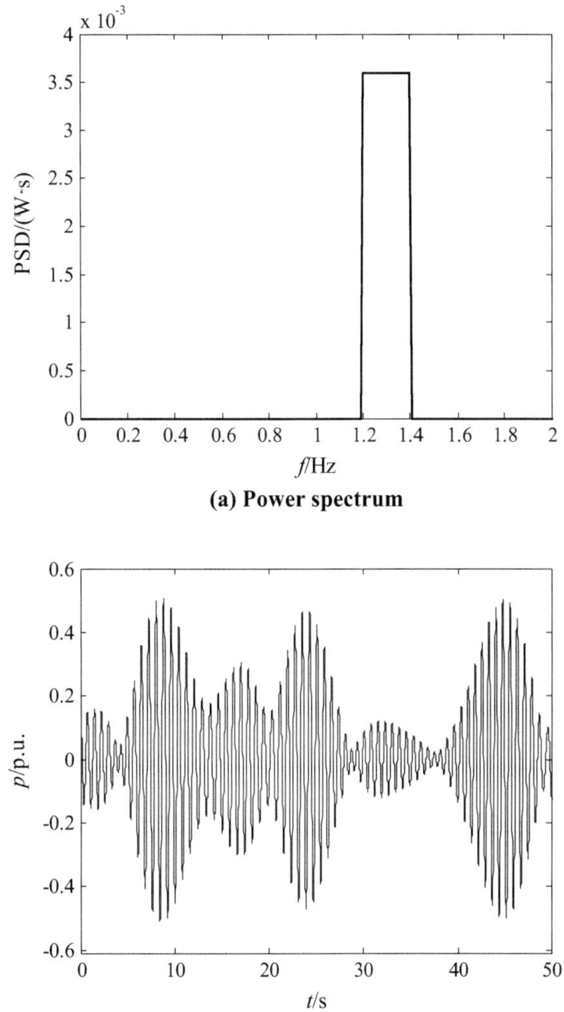

(a) Power spectrum

(b) Time-domain sequence

Fig. 5.41 Stochastic disturbance power spectrum and time domain sequence

phenomenon. In addition, in the relative power angle power spectrum of generator G3, the amplitude of the Mode 1 frequency component is larger than that of the Mode 2 frequency component. And at this moment, if attention is only paid to the power angle oscillation of the generator G3, it is very likely to cause misjudgment that Mode 1 is dominant while ignoring the real danger from Mode 2.

Comparing the initiation conditions of half-frequency internal resonance under different situations, it can be seen that under a special forced oscillation, if the system oscillation frequencies of Mode 1 and Modc 2 do not follow the ratio of 1:2, no half-frequency internal resonance can occur. Under a narrow-band stochastic disturbance, when $f_1 \in \left[\frac{1}{2}f_2 - \frac{1}{4}\Delta f,\ \frac{1}{2}f_2 + \frac{1}{4}\Delta f\right]$ is satisfied, Mode 1 can be excited.

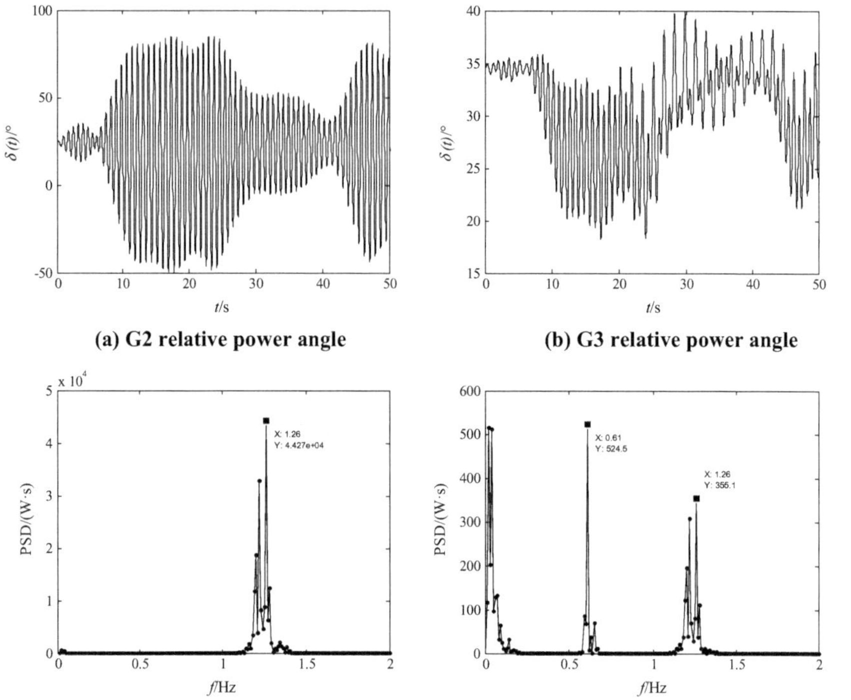

(a) G2 relative power angle **(b) G3 relative power angle**

(c) G2 relative power angle power spectrum **(d) G3 relative power angle power spectrum**

Fig. 5.42 Generator relative power angle power spectrum

That is, the phenomenon of half-frequency internal resonance in the power system can be more easily excited under the narrow-band stochastic disturbance.

References

1. Ju, Ping, Y.F. Liu, and Y.S. Xue, et al. 2017. Research prospects for stochastic dynamics of power system. *Automation of Electric Power Systems* 41(1): 1–8. (in Chinese).
2. Zhu, W.Q. 1998. *Stochastic Vibration*. Beijing: Science Press. (in Chinese).
3. Zhu, W.Q. 2003. *Nonlinear Stochastic Dynamics and Control: Hamiltonian Theoretical System Framework*. Beijing: Science Press. (in Chinese).
4. Tang, Y. 2006. Fundamental theory of forced power oscillation in power system. *Power System Technology* 30 (10): 29–33. (in Chinese).
5. Tang, Y. 1995. The analysis of forced power oscillation in power system. *Power System Technology* 19 (12): 6–10. (in Chinese).
6. Magdy, M.A., and F. Coowar. 1990. Frequency domain analysis of power system forced oscillations. *IEEE Proceedings: Generation, Transmission and Distribution* 137 (4): 261–268.

7. Yu, Y.P., Y. Ming, and L. Chen. 2009. Analysis of forced power oscillation steady-state response properties in multi-machine power systems. *Automation of Electric Power Systems* 33 (22): 5–9. (in Chinese).
8. Yu, Y.P., Y. Ming, and L. Chen, et al. 2010. Analysis of forced power oscillation caused by continuous cyclical load disturbances. *Automation of Electric Power Systems* 34 (6): 7–11. (in Chinese).
9. Chen, L., Y. Ming, and Y.P. Chen. 2014. Study on internal resonance in power system low frequency oscillation. *Power System Technology* 36 (1): 160–167. (in Chinese).
10. Byerly, R.T., R.J. Bennon, and D.E. Sherman. 1982. Eigenvalue analysis of synchronizing power flow oscillations in large electric power systems. *IEEE Transactions on Power Apparatus and Systems*, PAS-101(1): 235–243.
11. Jafarian, M., and A.M. Ranjbar. 2013. Interaction of the dynamics of doubly fed wind generators with power system electromechanical oscillations. *IET Renewable Power Generation* 7 (2): 89–97.
12. Chi, Su, Hu Weihao, Zhe Chen, et al. 2013. Mitigation of power system oscillation caused by wind power fluctuation. *IET Renewable Power Generation* 7 (6): 639–651.
13. Brownlees, S., D. Flynn, B. Fox, et al. 2007. The impact of wind farm power oscillations on the Irish power system. *IEEE Power Tech, Lausanne* 1–5: 195–200.
14. Newland, D.E. 1993. An introduction to random vibrations: spectral & wavelet analysis. 3rd ed. New York, USA: Longman Inc.

Chapter 6
Stochastic Dynamic Security of Power System

Abstract Traditionally, power system security means the ability of a power system to withstand sudden disturbances. To be specific, under dynamic conditions, the power system can withstand sudden disturbances and continuously provide users with the power and electric energy. It should be noted that the impact of stochastic property was seldom considered on the traditional power system security. Taking into account the stochastic property of the power system, the meaning of security needs to be expanded. Firstly, the disturbance includes not only "sudden disturbance" (e.g., a sudden short circuit), but also "continuous disturbances" (e.g., a stochastic disturbance). Secondly, the security assessment is not a deterministic problem, but a stochastic one. An index of security is not a deterministic one, but a probability one. The stochastic security of a power system in this chapter refers to the ability of a power system to withstand various disturbances. The disturbances described here include both traditional sudden disturbances and stochastic continuous disturbances. Essentially, the security requires the operating status and certain indexes of the system are maintained within certain ranges. To be specific, the system should be operated in an reliable area, such as maintaining the grid frequency and voltage within the normal level. In this chapter, "bound fluctuation region" which means the reliable range of the system under stochastic disturbances in engineering, is proposed and investigated. However, the problem of "dimensionality disaster" may occur if the states of the system are directly analyzed for the bound fluctuation region. In this chapter, the studies based on the system energy function are carried out for the bound fluctuation region, in which the high-dimension system state vector is simplified into a one-dimension value (i.e., system energy). Moreover, a quantitative analysis method is carried out for the bound fluctuation region.

6.1 Introduction

As renewable energy sources such as wind and solar power, as well as novel power loads like electric vehicles, become more and more common, stochastic power sources and loads take up a great portion of the power grid. Furthermore, the

P. Ju, *Stochastic Dynamics of Power Systems*, Power Systems,
https://doi.org/10.1007/978-981-13-1816-0_6

intensity of stochastic disturbances increases, which seriously threatens the security and stability of the power grid. Consequently, it is necessary to conduct a detailed assessment for the security and stability of the stochastic power system under any underlying dangers, and necessary measures should be defined to improve the security and stability margins and to ensure the system is operated within them.

Stochastic dynamic system operation trajectories are random. A stochastic disturbance with the same intensity may excite completely opposite results. When the stochastic disturbances are added into the power system, stable systems may lose the stability. Therefore, the analysis of a stochastic system is unlike that of a deterministic system, so it is impossible to analyze the stability of a power system subjected to stochastic disturbances just based on one time-domain simulation under a specific disturbance. Likewise, it is impossible to measure the transient stability margin of a stochastic system using traditional critical clearing times of faults, stability margins and other indexes. Therefore, the stability of stochastic power systems should be described via a series of probability indexes.

In recent years, series of research on the security and stability of power systems subjected to stochastic disturbances have been carried out. There are two main types of the analysis methods. The first one is the probability-based method, in which the Monte Carlo simulation is the most representative which calculates, counts and analyzes the probability distribution function of system stability. However, the Monte Carlo simulation still regards stochastic systems as deterministic systems in each sampling. Furthermore, the Monte Carlo simulation requires a huge amount of computation, especially when studying the influence of system parameter changes on stability. Monte Carlo simulation needs to be performed for different parameter values, but the amount of calculation is too large to perform.

The application of stochastic dynamics theory to studying the security and stability of stochastic power systems is mainly based on the stochastic differential algebraic equations (SDAE) models, numerical solutions, etc., and it is still in its infancy. For that reason, there is an urgent need to find more efficient and reliable methods for the stochastic analysis.

6.2 Description of Stochastic Dynamic Security of Power System

Under the stochastic disturbance, the state variables of a power system such as the voltage, current and frequency become stochastic variables which are uncertain. For high-dimension power systems, their states can be described by stochastic vectors. From the point of view of the safe operation of the power system, it is necessary to know whether the state of the system at this point can be kept within a security field acceptable to the project. The acceptable range security of the system state under the stochastic disturbance is defined as the bound fluctuation region.

Under stochastic disturbances, the system running state $\boldsymbol{X}(t)$ becomes a stochastic vector at time t. Therefore, whether $\boldsymbol{X}(t)$ can stay within the bound fluctuation region is a stochastic event and needs to be described by probability. Under stochastic disturbance, the probability of a system state always being within the bound fluctuation region at t time is an intra-region probability, which can be expressed as:

$$R(t|\boldsymbol{X}_0) = P\{\boldsymbol{X}(s) \in \Omega, s \in (0,t]|\boldsymbol{X}(0) = \boldsymbol{X}_0 \in \Omega\} \tag{6.2.1}$$

In such equation: $\boldsymbol{X}_0$ denotes the initial state of the system, Ω denotes the system bound fluctuation region.

It is feasible to analytically calculate the intra-region probability (6.2.1) in an one machine infinite bus system with only two-dimension states. However, the calculation becomes burdensome in large-scale power systems, due to the high-dimension operating states. It is easy to understand that the operating state with a higher system energy value is further from the equilibrium point. In other words, the system energy can be regarded as the 'distance' of an operating state from the equilibrium point. Hence, it is reasonable that an operating state is in a predefined region (i.e., the 'distance' of the operating state from equilibrium point is limited) when the system energy value is lower than a preset value. Based on the above analysis, the intra-region probability can be expressed by the following form:

$$P(t|H_0) = P\{H(\tau) < H_B, \tau \in (0,t]|H(0) = H_0 < H_B\} \tag{6.2.2}$$

where $H(t)$ denotes the system energy along with the time, H_B denotes the predefined energy boundary, and $P(t|H_0)$ denotes the probability of $H(t)$ being lower than H_B.

6.3 Assessment Method for Stochastic Dynamic Security of Power System

For a linear stochastic system, if the disturbance is Gaussian white noise, Itô stochastic differential equations can be used to work out the matrix of the system solution and exact solutions to correlation functions. However, for non-Gaussian disturbance linear systems or nonlinear stochastic systems, exact solutions can not be obtained. Only Monte Carlo simulation can be used to calculate a large number of samples, and then the statistical analysis is performed for obtaining the response. Due to a large amount of time consuming, the analysis of stochastic systems is very difficult. In recent years, people have gradually developed some approximate solutions, such as the equivalent linear approximation method, ignoring the high-order cumulants truncation method, the equivalent nonlinear system method, and the stochastic averaging method (SAM).

6.3.1 Stochastic Averaging Method

The stochastic averaging method was first proposed by Stratonovich [1] and then mathematically proven by Khasminskii [2]. The SAM can be used in two cases: the stochastic averaging of amplitude envelope and stochastic averaging of energy envelope. SAM generally consists of two steps: the first step is to approximate a non-white Gaussian noise disturbance with a white Gaussian noise disturbance, turning the system response to a Markov-diffusion process; the second step is time averaging. In fact, as long as there are one or more slow-changing response processes in a stochastic dynamic system, the SAM can be applied.

Zhu Weiqiu combined the quasi Hamiltonian system theory with the SAM [3, 4], first put forward the quasi Hamiltonian system SAM theory, and established a complete set of quasi-Hamiltonian-system-based nonlinear stochastic dynamical analysis methods. This theory is proven to be a powerful tool for analyzing strong nonlinear stochastic dynamical systems with multiple degrees of freedom. It provides a good theoretical tool for the analysis of stochastic transient stability of power systems.

It is well-known that the classical Hamiltonian system is autonomous and conservative. However, many engineering systems, including the power system, are systems having energy dissipation and exchange with the outside world, not like the classical Hamiltonian system, which, to a large extent, limits the application of the Hamiltonian system theory. A Quasi-Hamiltonian system theory extends to a wider system range including dissipative, control and considering stochastic disturbances, so the problem can be solved better.

For the power system, the stochastic disturbance source occupies a capacity in the total system capacity by a relatively small ratio, which can meet the condition that the energy exchanged with the outside system is smaller than the total energy, so this system and can be regarded as the quasi Hamiltonian system. By contrast, in the stochastic power system the generator rotor angle and rotation speed change rapidly, which can be regarded as fast variables. However, the transient energy (Hamilton function) changes slowly and can be regarded as a slow variable.

In the following, the quasi Hamiltonian system SAM and its application in power systems are focused on. Under the quasi Hamiltonian system, an n-degree-of-freedom nonlinear stochastic dynamical system can be expressed as:

$$\begin{cases} \dot{q}_i = \frac{\partial H'}{\partial p_i} \\ \dot{p}_i = -\frac{\partial H'}{\partial q_i} - \varepsilon \sum_{j=1}^{n} c_{ij}(Q,P)\frac{\partial H'}{\partial p_j} + \varepsilon^{\frac{1}{2}} \sum_{k=1}^{m} f_{ik}(Q,P)W_k(t) + u_i(Q,P) \end{cases} \tag{6.3.1}$$

where $i, j = 1, 2, \ldots, n$. q_i, p_i are the displacements and momenta of the system, with the vector forms recorded as Q, P, $H'(Q,P)$ is the system Hamilton function. ε is a small parameter, $c_{ij}(Q,P)$ is the system damping coefficient, and W_k is

Gaussian white noise, the correlation functions of which are: $E[W_l(t)W_s(t+\tau)] = 2\pi K_{ls}\delta(\tau)$, $l, s = 1, 2, \ldots, m$. $f_{ik}(Q, P)$ is the action coefficient of the stochastic disturbance W_k upon the state quantity, $u_i(Q, P)$ is the additional control. If the damping force and the stochastic disturbance are relatively small in a cycle magnitude of time, and the difference between the energy input by the stochastic disturbance into the system and the energy consumed by the damping is smaller than the total energy of the system itself, this system is called the quasi Hamiltonian system.

The basic principle of the SAM of the quasi Hamiltonian system is as below. According to the ergodic assumption of classical statistical mechanics, the state of the quasi Hamiltonian system can be assumed to be equal-probability distributed on the equal energy surface. Furthermore, substituting the time average with the space average, averaging the fast variables, keeping the slow variables, calculating the drift coefficient and diffusion coefficient of the average system, and finally getting the Itô average diffusion equation about slow variables. Since the average system contains only slow variables, its dimension is generally much smaller than the original system dimension. Due to the most mature research on diffusion equation in the stochastic dynamic theory, the existing theoretical methods suitable for diffusion process can all be used to analyze the average system. The SAM theory of quasi Hamiltonian system applies the SAM to the analysis of the quasi Hamiltonian system, which not only simplifies the analysis but also reduces the computational complexity. This method has been successfully applied to the analysis of many nonlinear stochastic systems, results to which are in good agreement with the numerical simulation method.

A detailed derivation process of the SAM can be found in the literature [5]. For the quasi Hamiltonian system described in Eq. (6.3.1), it is assumed for simplicity that the additive control $u_i(Q, P)$ is zero, and the Stratonovich differential equation model can be established first and then converted to Itô stochastic differential equation, then an Itô stochastic differential equation about $H(t)$ can be obtained according to Itô differential rules. Since $H(t)$ is a slow-changing process while $q_1, q_2, \ldots, q_n, p_2, \ldots, p_n$ is a fast one, the Khasminskii principle of stochastic averaging can be applied, and the Markov diffusion process can be used to approximate $H(t)$, dominated by the Itô stochastic differential equation as follows:

$$\mathrm{d}H = m(H)\mathrm{d}t + \sigma(H)\mathrm{d}B(t) \tag{6.3.2}$$

where $B(t)$ is a unit Wiener process, and the drift coefficient $m(H)$ and the diffusion coefficient $\sigma(H)$ can be obtained from the time average:

$$m(H) = \varepsilon\left\langle -\sum_{i,j=1}^{n} m_{ij}\frac{\partial H}{\partial p_i}\frac{\partial H}{\partial q_k} + \sum_{i,j=1}^{n}\sum_{l,s=1}^{m}\pi K_{ls} g_{il} g_{is}\frac{\partial^2 H}{\partial p_i \partial p_j}\right\rangle_t \tag{6.3.3}$$

$$\sigma^2(H) = \varepsilon \cdot 2\pi \left\langle \sum_{i,j=1}^{n} \sum_{l,s=1}^{m} K_{ls} g_{il} g_{jl} \frac{\partial H}{\partial p_i} \frac{\partial H}{\partial p_j} \mathrm{d}B_l(t) \right\rangle_t \tag{6.3.4}$$

The fully non-integrable Hamiltonian system traverses across a $(2n-1)$-dimension isoenergetic surface, such that the time average can be replaced by the space average of the fast-changing variables $q_1, q_2, \ldots, q_n, p_2, \ldots, p_n$. Specifically, the time average of a function $F(q,p)$ can be calculated as follows:

$$\langle F(q,p) \rangle_t = \frac{1}{T(H)} \int_{\Omega} F(q,p) \left(\frac{\partial H}{\partial p_1} \right)^{-1} \mathrm{d}q_1 \mathrm{d}q_2 \ldots \mathrm{d}q_n \mathrm{d}p_2 \ldots \mathrm{d}p_n \tag{6.3.5}$$

$$T(H) = \int_{\Omega} \left(\frac{\partial H}{\partial p_1} \right)^{-1} \mathrm{d}q_1 \mathrm{d}q_2 \ldots \mathrm{d}q_n \mathrm{d}p_2 \ldots \mathrm{d}p_n \tag{6.3.6}$$

And the integral field Ω is defined as $H(q_1, q_2, \ldots, q_n, p_1 = 0, p_2, \ldots, p_n) \leq H$. After the average, a drift coefficient $m(H)$ and a diffusion coefficient $\sigma(H)$ are obtained. By solving the Fokker-Plank-Kolmogorov (FPK) equation corresponding to the averaged diffusion Eq. (6.3.2):

$$\frac{\partial}{\partial t} p + \frac{\partial}{\partial H} (m(H) p) - \frac{1}{2} \frac{\partial^2}{\partial H^2} \left(\sigma^2(H) p \right) = 0 \tag{6.3.7}$$

And according to the corresponding initial condition: $p(H, 0 | H_0) = \delta(H - H_0)$, we can find the transition probability density of H: $p(H, t | H_0, t_0)$.

6.3.2 EEAC Method

When the quasi Hamiltonian system SAM is used to analyze a stochastic power system, the expression of Hamilton function H is very complicated if the power system is large-scale with many generators. Likewise, the time averaging operation is complicated for Eqs. (6.3.5) and (6.3.6). In order to simplify the problem and reduce the computational cost, one can adopt extended equal area criterion (EEAC), which is commonly used in power system transient stability analysis, to make the multi-machine system equivalent to a two-machine equivalent.

The EEAC method has the following three forms: static EEAC, dynamic EEAC and integrated EEAC. Among them, the static EEAC has a higher accuracy when the two-group characteristics of the power system are good, but a lower one when the two-group characteristics are weak. The dynamic EEAC method and the integrated EEAC method are improved on this basis. The dynamic EEAC method improves the parameters of the two-machine equivalent system segment by segment according to a predicted trajectory, in order to improve the accuracy. And, the integrated EEAC method improves the analysis accuracy through keeping stability and reducing the dimension of the trajectory mapping. The literature [6] describes the principle in detail.

Hereinafter is a brief introduction of the static EEAC method, assuming that a classic model of multi-machine disturbed system is:

$$\begin{cases} \dot{\delta}_i = (\omega_i - 1)\omega_N \\ M_i\dot{\omega}_i = P_{mi}\left(1 + \sum_{k=1}^{m} f_{ik}W_k\right) - P_{ei} - D_i\omega_i \end{cases} \tag{6.3.8}$$

where δ_i, ω_i are the generator power angle and rotation speed, respectively; ω_N is the synchronous speed; W_k denotes a Gaussian white noise with its intensity being ζ_k; f_{ik} denotes the noise coefficient; D_i denotes the damping coefficient given that $D_i/M_i = k$; and P_{mi} and P_{ei} denote the generator mechanical power and the output electromagnetic power, respectively.

According to the EEAC method, the instability of a multi-machine system can be regarded as a group formed by one or several generators that are out of synchronization with the rest of the system. The severely disturbed group is referred to as a forward (S) group, and the remaining groups are backward (A) groups. By using COI (center of inertia) coordinates, one can deduce the disturbed two-machine equivalent model of the stochastic disturbance power system:

$$\begin{cases} \dot{\delta}_{SA} = \omega_{SA} \cdot \omega_N \\ M\dot{\omega}_{SA} = P_{mSA} - (P_c + P_{\max}\sin(\delta_{SA} - \gamma)) - Mk\omega_{SA} + \sum_{k=1}^{m} Ml_kW_k \end{cases} \tag{6.3.9}$$

where definitions of $l_k = \frac{1}{M_S}\sum_{i\in S} P_{mi}f_{ik} - \frac{1}{M_A}\sum_{i\in A} P_{mi}f_{ik}$, $M, P_{\max}, P_{eSA}, P_c, \gamma$ and other parameters can be found in the literature [6]. If the system has better two-group characteristics, the two-machine equivalent system is an ideal Hamiltonian system without considering the effect of non-Hamiltonian factors such as damping and stochastic disturbances. The Hamiltonian function is

$$\begin{aligned} H = &\frac{1}{2}M\omega_N\omega_{SA}^2 - (P_{mSA} - P_c)(\delta_{SA} - \delta_{SA0}) \\ &- P_{\max}\left(\cos(\delta_{SA} - \gamma) - \cos\left(\delta_{SA,0} - \gamma\right)\right) \end{aligned} \tag{6.3.10}$$

After considering the influence of damping and stochastic disturbances, the quasi Hamilton system theory can be applied to describe the disturbed two-machine equivalent system, which can be rewritten as:

$$\begin{cases} \dot{\delta}_{SA} &= \frac{1}{M}\frac{\partial H}{\partial \omega_{SA}} \\ \dot{\omega}_{SA} &= -\frac{1}{M}\frac{\partial H}{\partial \delta_{SA}} - \frac{k}{M\omega_N}\frac{\partial H}{\partial \omega_{SA}} + \sum_{k=1}^{m} l_i W_i \end{cases} \tag{6.3.11}$$

The SAM can be used to analyze the Hamiltonian energy diffusion process of the quasi Hamiltonian system (6.3.11), so as to study the security and stability of the original multi-machine disturbed system.

When the two-group characteristics are better, the equivalent system has a better Hamiltonian property. If not, the equivalent system shows a more obvious non-Hamiltonian property. So, for stochastic power systems with better two-group characteristics, the SAM can be directly applied based on the quasi Hamiltonian system, which is described in Sect. 6.4. For stochastic power systems with weaker two-group characteristics, the analysis accuracy decreases.

6.3.3 Moment Analysis Method

When the intensity of stochastic disturbances in the power system is not large, the system can be described as an approximate linearization model. The moment analysis method of the stochastic linear system can be applied in turn. This method is not limited to the complexity of the system model. Even a complex system model can be linearized to obtain its linear model subjected to stochastic disturbance and then be analyzed.

The deterministic model of a power system is usually described through a set of differential algebraic equations (DAEs) as follows:

$$\begin{cases} \mathrm{d}\boldsymbol{X}(t) = f[\boldsymbol{X}(t), \boldsymbol{Y}(t)]\mathrm{d}t \\ \boldsymbol{0} = g[\boldsymbol{X}(t), \boldsymbol{Y}(t)] \end{cases} \tag{6.3.12}$$

where $\boldsymbol{X}(t)$ denotes the vector of state variables; $\boldsymbol{Y}(t)$ denotes the vector of algebraic variables; $\boldsymbol{f}$ denotes the vector of nonlinear functions in differential equations; $\boldsymbol{g}$ denotes the vector of nonlinear functions in algebraic equations; and $\boldsymbol{O}$ denotes a null matrix.

After considering the stochastic disturbance, continuous-time stochastic processes are introduced into the DAEs (6.3.12). Then, a set of SDAEs, which can be used to describe power systems under stochastic disturbance, is built as follows:

$$\begin{cases} \mathrm{d}\boldsymbol{X}(t) = f[\boldsymbol{X}(t), \boldsymbol{Y}(t)]\mathrm{d}t + \boldsymbol{K}\mathrm{d}\boldsymbol{B}(t) \\ \boldsymbol{0} = g[\boldsymbol{X}(t), \boldsymbol{Y}(t)] \end{cases} \tag{6.3.13}$$

where $\boldsymbol{K}\mathrm{d}\boldsymbol{B}(t)$ denotes the vector of stochastic disturbance; $\boldsymbol{B}(t)$ denotes a vector whose elements are independent standard Wiener processes; and $\boldsymbol{K}$ denotes the covariance matrix of stochastic disturbance, which means the intensity and correlation of stochastic disturbance.

Based on the equilibrium point (i.e., the steady state) of the nonlinear model (6.3.12), one can linearize (6.3.13) around it. The linearized stochastic model of a power system under stochastic disturbance can be expressed as follows:

$$\begin{bmatrix} \mathrm{d}\Delta\boldsymbol{X}(t) \\ \boldsymbol{0} \end{bmatrix} = \begin{bmatrix} \boldsymbol{f}_X & \boldsymbol{f}_Y \\ \boldsymbol{g}_X & \boldsymbol{g}_Y \end{bmatrix} \begin{bmatrix} \Delta\boldsymbol{X}(t) \\ \Delta\boldsymbol{Y}(t) \end{bmatrix} \mathrm{d}t + \begin{bmatrix} \boldsymbol{K}\mathrm{d}\boldsymbol{B}(t) \\ \boldsymbol{0} \end{bmatrix} \tag{6.3.14}$$

where $\Delta\boldsymbol{X}(t) = \boldsymbol{X}(t) - \boldsymbol{X}(0)$ and $\Delta\boldsymbol{Y}(t) = \boldsymbol{Y}(t) - \boldsymbol{Y}(0)$; $\boldsymbol{X}(0)$ and $\boldsymbol{Y}(0)$ denote the initial point, which is assumed as the equilibrium point of the system in this study; $\boldsymbol{f}_X$ and $\boldsymbol{f}_Y$ are the Jacobian matrixes of $\boldsymbol{f}$ with respect to $\boldsymbol{X}$ and $\boldsymbol{Y}$, which can be obtained by differential derivatives. $\boldsymbol{g}_X$ and $\boldsymbol{g}_Y$ are the Jacobian matrixes of $\boldsymbol{g}$ with respect to $\boldsymbol{X}$ and $\boldsymbol{Y}$, which can be taken straight from the power flow Jacobian matrix.

Eliminating the algebraic equation in (6.3.14), one can obtain the linearized stochastic model as a set of linearized stochastic differential equations (LSDEs), which is as follows:

$$\mathrm{d}\Delta\boldsymbol{X}(t) = \boldsymbol{A}\Delta\boldsymbol{X}(t)\mathrm{d}t + \boldsymbol{K}\mathrm{d}\boldsymbol{B}(t) \tag{6.3.15}$$

where $\boldsymbol{A}$ denotes the state matrix of the linearized system and equals to $\boldsymbol{f}_X - \boldsymbol{f}_Y(\boldsymbol{g}_Y)^{-1}\boldsymbol{g}_X$.

In the following, the explicit formulation of solution for LSDEs (6.3.15) is presented. Furthermore, the mean and covariance of system states over time, which are closely related to the characteristics of systems states, are analytically solved.

Before giving the explicit formulation of the solution to LSDEs (6.3.15), an exponential function of a general nth-order square matrix $\boldsymbol{M}$ is firstly defined as follows:

$$\exp(\boldsymbol{M}) = \sum_{n=0}^{+\infty} (\boldsymbol{M}^n/n!) \tag{6.3.16}$$

where $n!$ denotes the factorial of a non-negative integer n.

Multiplying both sides of LSDEs (6.3.15) by $\exp(-\boldsymbol{A}t)$, one obtains:

$$\exp(-\boldsymbol{A}t)\mathrm{d}\Delta\boldsymbol{X}(t) = \exp(-\boldsymbol{A}t)\boldsymbol{A}\Delta\boldsymbol{X}(t)\mathrm{d}t + \exp(-\boldsymbol{A}t)\boldsymbol{K}\mathrm{d}\boldsymbol{B}(t) \tag{6.3.17}$$

By the matrix calculus, (6.3.17) can be expressed as:

$$\mathrm{d}[\exp(-\boldsymbol{A}t)\Delta\boldsymbol{X}(t)] = \exp(-\boldsymbol{A}t)\boldsymbol{K}\mathrm{d}\boldsymbol{B}(t) \tag{6.3.18}$$

Let s denote t in (6.3.18), and then the integral of (6.3.18) over the time interval $[t_0, t]$ is deduced as follows:

$$\exp(-\boldsymbol{A}t)\Delta\boldsymbol{X}(t) - \exp(-\boldsymbol{A}t_0)\Delta\boldsymbol{X}(t_0) = \int_{t_0}^{t} \exp(-\boldsymbol{A}s)\boldsymbol{K}\mathrm{d}\boldsymbol{B}(s). \tag{6.3.19}$$

Multiplying both sides of (6.3.19) by $\exp(\boldsymbol{A}t)$, one obtains:

$$\Delta\boldsymbol{X}(t) = \exp[\boldsymbol{A}(t - t_0)]\Delta\boldsymbol{X}(t_0) + \int_{t_0}^{t} \exp[\boldsymbol{A}(t - s)]\boldsymbol{K}\mathrm{d}\boldsymbol{B}(s). \tag{6.3.20}$$

Equation (6.3.20) is the explicit formulation of the solution to LSDEs (6.3.15). It was proven that $\int_{t_0}^{t} \exp[\boldsymbol{A}(t - s)]\boldsymbol{K}\mathrm{d}\boldsymbol{B}(s)$ follows Gaussian distributions, and thus system states $\Delta\boldsymbol{X}(t)$ follow Gaussian distributions. For the Gaussian distribution, the mean and variance are two key factors.

According to the definition of mean, the system states mean under stochastic disturbance are expressed as follows:

$$\begin{aligned} E[\Delta\boldsymbol{X}(t)] &= E\{\exp[\boldsymbol{A}(t - t_0)]\Delta\boldsymbol{X}(t_0) + \int_{t_0}^{t} \exp[\boldsymbol{A}(t - s)]\boldsymbol{K}\mathrm{d}\boldsymbol{B}(s)\} \\ &= E\{\exp[\boldsymbol{A}(t - t_0)]\Delta\boldsymbol{X}(t_0)\} + E\{\int_{t_0}^{t} \exp[\boldsymbol{A}(t - s)]\boldsymbol{K}\mathrm{d}\boldsymbol{B}(s)\}. \end{aligned} \tag{6.3.21}$$

By the definition of Itô integral, the following formulation can be deduced from (6.3.21),

$$\begin{aligned} E[\Delta\boldsymbol{X}(t)] &= E\{\exp[\boldsymbol{A}(t - t_0)]\Delta\boldsymbol{X}(t_0)\} + E\left\langle \lim_{\lambda\to 0} \sum\nolimits_{i=0}^{k-1} \{\exp[\boldsymbol{A}(t - t_i)]\boldsymbol{K}[\boldsymbol{B}(t_{i+1}) - \boldsymbol{B}(t_i)]\} \right\rangle \\ &= E\{\exp[\boldsymbol{A}(t - t_0)]\Delta\boldsymbol{X}(t_0)\} + \lim_{\lambda\to 0} \sum\nolimits_{i=0}^{k-1} \{\exp[\boldsymbol{A}(t - t_i)]\boldsymbol{K}E[\boldsymbol{B}(t_{i+1}) - \boldsymbol{B}(t_i)]\} \\ &= E\{\exp[\boldsymbol{A}(t - t_0)]\Delta\boldsymbol{X}(t_0)\} \end{aligned} \tag{6.3.22}$$

where $t_k = t$; λ is equal to the maximum value of $t_{i+1} - t_i$ $(i = 0, 1, \ldots, k - 1)$; $E[\boldsymbol{B}(t_{i+1}) - \boldsymbol{B}(t_i)] = 0$, because of the characteristic of Wiener process.

The parameters $\boldsymbol{A}$ and initial system states $\Delta\boldsymbol{X}(t_0)$ are assumed without randomness. Thus, one obtains:

$$E[\Delta \boldsymbol{X}(t)] = \exp[\boldsymbol{A}(t - t_0)]\Delta \boldsymbol{X}(t_0). \tag{6.3.23}$$

Equation (6.3.23) is the explicit formulation of the solution to LSDEs (6.3.15) system states mean.

The system states variance is an important factor, which indirectly indicates how far system states will leave the equilibrium point under stochastic disturbance. According to the definition of the covariance matrix, the system states covariance matrix can be expressed as follows:

$$D[\Delta \boldsymbol{X}(t)] = E\langle\{\Delta \boldsymbol{X}(t) - E[\Delta \boldsymbol{X}(t)]\}\{\Delta \boldsymbol{X}(t) - E[\Delta \boldsymbol{X}(t)]\}^T\rangle \tag{6.3.24}$$

Substituting $\Delta \boldsymbol{X}(t)$ (6.3.20) and its mean (6.3.23) into the covariance matrix (6.3.24), the following formulation can be deduced,

$$D[\Delta \boldsymbol{X}(t)] = E\left\langle \{\int_{t_0}^{t} \exp[\boldsymbol{A}(t-s)]\boldsymbol{K}\mathrm{d}\boldsymbol{B}(s)\}\{\int_{t_0}^{t} \exp[\boldsymbol{A}(t-s)]\boldsymbol{K}\mathrm{d}\boldsymbol{B}(s)\}^T \right\rangle \tag{6.3.25}$$

By the definition of Itô integral, the following equation can be further deduced from (6.3.25),

$$\begin{aligned}
D[\Delta \boldsymbol{X}(t)] &= E\left\langle \{\lim_{\lambda\to 0} \textstyle\sum_{i=0}^{k-1} \exp[\boldsymbol{A}(t-t_i)]\boldsymbol{K}[\boldsymbol{B}(t_{i+1}) - \boldsymbol{B}(t_i)]\} \right. \\
&= E\left\langle \lim_{\lambda\to 0} \sum_{0\le i,j\le k-1} \{\exp[\boldsymbol{A}(t-t_i)]\boldsymbol{K}[\boldsymbol{B}(t_{i+1}) - \boldsymbol{B}(t_i)] \right. \\
&\quad \times [\boldsymbol{B}(t_{j+1}) - \boldsymbol{B}(t_j)]^T\boldsymbol{K}^T\exp[\boldsymbol{A}^T(t-t_j)]\}\rangle \\
&= \lim_{\lambda\to 0} \sum_{0\le i,j\le k-1} \langle\exp[\boldsymbol{A}(t-t_i)]\boldsymbol{K}E\{[\boldsymbol{B}(t_{i+1}) - \boldsymbol{B}(t_i)] \\
&\quad \times [\boldsymbol{B}(t_{j+1}) - \boldsymbol{B}(t_j)]^T\}\boldsymbol{K}^T\exp[\boldsymbol{A}^T(t-t_j)]\rangle \\
&= \lim_{\lambda\to 0} \sum_{\substack{0\le i,j\le k-1,\\ \text{and } i=j}} \langle\exp[\boldsymbol{A}(t-t_i)]\boldsymbol{K}E\{[\boldsymbol{B}(t_{i+1}) - \boldsymbol{B}(t_i)] \\
&\quad \times [\boldsymbol{B}(t_{j+1}) - \boldsymbol{B}(t_j)]^T\}\boldsymbol{K}^T\exp[\boldsymbol{A}^T(t-t_j)]\rangle \\
&\quad + \lim_{\lambda\to 0} \sum_{\substack{0\le i,j\le k-1,\\ \text{and } i\ne j}} \langle\exp[\boldsymbol{A}(t-t_i)]\boldsymbol{K}E\{[\boldsymbol{B}(t_{i+1}) - \boldsymbol{B}(t_i)] \\
&\quad \times [\boldsymbol{B}(t_{j+1}) - \boldsymbol{B}(t_j)]^T\}\boldsymbol{K}^T\exp[\boldsymbol{A}^T(t-t_j)]\rangle
\end{aligned} \tag{6.3.26}$$

where $t_k = t$; λ is equal to the maximum value of $t_{i+1} - t_i$ ($i = 0, 1, \ldots, k - 1$) and $t_{j+1} - t_j$ ($j = 0, 1, \ldots, k - 1$).

By the characteristic of Wiener processes, one obtains:

$$E\{[\boldsymbol{B}(t_{i+1}) - \boldsymbol{B}(t_i)][\boldsymbol{B}(t_{j+1}) - \boldsymbol{B}(t_j)]^T\} = \begin{cases} t_{i+1} - t_i, & if\ i = j \\ 0, & if\ i \neq j \end{cases} \tag{6.3.27}$$

Substituting (6.3.27) into the covariance matrix (6.3.26), one obtains:

$$\begin{aligned} D[\Delta \boldsymbol{X}(t)] &= \lim_{\lambda \to 0} \sum_{i=0}^{k-1} \{\exp[\boldsymbol{A}(t - t_i)]\boldsymbol{K}\boldsymbol{K}^T \exp[\boldsymbol{A}^T(t - t_i)](t_{i+1} - t_i)\} \\ &= \int_{t_0}^{t} \exp[\boldsymbol{A}(t - s)]\boldsymbol{K}\boldsymbol{K}^T \exp[\boldsymbol{A}^T(t - s)]\mathrm{d}s \end{aligned} \tag{6.3.28}$$

Equation (6.3.28) is the formulation of the solution to LSDEs (6.3.15) system states covariance matrix.

6.4 Dynamic Security of Power System under Small Stochastic Disturbance

6.4.1 Statistics of State Fluctuation in Power System

When a stochastic disturbance is applied to the system, the states of the system oscillate. It is known that the forced oscillation is usually excited by deterministic continuous disturbances, whose frequencies are close or equal to system natural frequencies. In the conventional deterministic analysis approaches of forced oscillation, disturbances are assumed with deterministic characteristics in the frequency domain, and thus the system states oscillate along with deterministic characteristics in the frequency domain. With the increasing randomness in power system, these deterministic methods can face new problems, due to no consideration of the random nature. For example, the intermittent renewable energy and plug-in electric vehicles are introducing a stochastic disturbance into the power system, whose frequencies and amplitudes are stochastic. To show the forced oscillation under stochastic disturbance, Fig. 6.1 gives five rotor angle trajectories of a single-machine infinite-bus system (SMIB) under stochastic disturbance. In this SMIB system, all real parts of system eigenvalues are negative, and the initial simulation value is set as the stable equilibrium point. It can be seen that the rotor angle oscillates with stochastic amplitudes over time, which cannot be revealed and explained by the conventional deterministic analysis approaches. In other words,

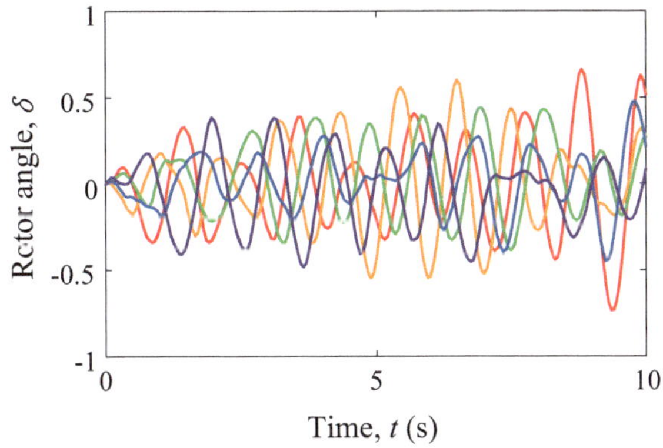

Fig. 6.1 Rotor angle oscillation trajectories of an SMIB system under stochastic disturbance

for the forced oscillation under stochastic disturbance, conventional deterministic methods without consideration of the random nature are incomplete. Thus, stochastic analytic methods which can offer a more comprehensive analysis are more desired.

With the increasing penetration of intermittent renewable energy and plug-in electric vehicles, stochastic disturbance becomes more and more considerable. In this study, for modeling stochastic disturbance, an approach similar to the existed study is adopted. Specifically, the stochastic disturbance is modeled as continuous-time stochastic processes, additive to the power system models. For example, when the stochastic disturbance is considered from the renewable generation, the mechanical power input of a wind turbine can be expressed as:

$$P_m = P_{m,0} + \Delta P_m \tag{6.4.1}$$

where P_m denotes the mechanical power including stochastic disturbance; $P_{m,0}$ denotes the low-frequency mechanical power, which is assumed to be constant and equal to the mean value of P_m; and ΔP_m denotes the high-frequency stochastic disturbance, which can be described by an additive continuous-time stochastic process.

Furthermore, the linearized stochastic differential equations for this system can be expressed as follows (more details can be found in Sect. 6.3.3):

$$\mathrm{d}\Delta \boldsymbol{X}(t) = \boldsymbol{A}\Delta \boldsymbol{X}(t)\mathrm{d}t + \boldsymbol{K}\mathrm{d}\boldsymbol{B}(t) \tag{6.4.2}$$

Based on the analyzed method in Sect. 6.3.3, the mean and covariance matrix of the system states in (6.4.10) can be expressed as:

$$\begin{cases} E[\Delta \boldsymbol{X}(t)] = \exp[\boldsymbol{A}(t-t_0)]\Delta \boldsymbol{X}(t_0) \\ D[\Delta \boldsymbol{X}(t)] = \int_{t_0}^{t} \exp[\boldsymbol{A}(t-s)]\boldsymbol{K}\boldsymbol{K}^T \exp[\boldsymbol{A}^T(t-s)]\mathrm{d}s \end{cases} \tag{6.4.3}$$

By utilizing (6.4.3), one can calculate the system states covariance matrix during the forced oscillation. However, the solution to covariance matrix (6.4.3) is not exactly analytic, which lend itself difficult to calculation, due to the existence of the integral. In the following, matrix operations are adopted to eliminate the integral in the covariance matrix (6.4.3), and then an analytic expression of the covariance matrix without the integral is deduced. To analyze the forced oscillation, the analysis system is assumed to be small-signal stable, which means all the eigenvalues exist and have negative real parts. By eigen decomposition, system state matrix $\boldsymbol{A}$ can be factorized as:

$$\boldsymbol{A}=\boldsymbol{P}\boldsymbol{\Lambda}\boldsymbol{P}^{-1} \tag{6.4.4}$$

where $\boldsymbol{\Lambda}$ denotes the eigenvalue matrix, and $\boldsymbol{\Lambda}$ = diag[λ_1, λ_2, …, λ_i, …]; and $\boldsymbol{P}$ denotes the similarity transformation from $\boldsymbol{A}$ to $\boldsymbol{\Lambda}$, whose ith column is the eigenvector.

Substituting (6.4.12) into exp[$\boldsymbol{A}(t-s)$], one obtains:

$$\exp[\boldsymbol{A}(t-s)]=\boldsymbol{P}\boldsymbol{\Gamma}\boldsymbol{P}^{-1} \tag{6.4.5}$$

where $\boldsymbol{\Gamma}$ = diag[exp($\lambda_1 t - \lambda_1 s$), exp($\lambda_2 t - \lambda_2 s$), …, exp($\lambda_i t - \lambda_i s$), …].

Incorporating exp[$\boldsymbol{A}(t-s)$] (6.4.13) into $D[\Delta\boldsymbol{X}(t)]$ in (6.4.3), one obtains:

$$D[\Delta\boldsymbol{X}(t)] = \int_{t_0}^{t} \boldsymbol{P}\boldsymbol{\Gamma}\boldsymbol{P}^{-1}\boldsymbol{K}\boldsymbol{K}^T(\boldsymbol{P}^{-1})^T\boldsymbol{\Gamma}^T\boldsymbol{P}^T\mathrm{d}s = \boldsymbol{P}[\int_{t_0}^{t} \boldsymbol{\Gamma}\boldsymbol{P}^{-1}\boldsymbol{K}\boldsymbol{K}^T(\boldsymbol{P}^{-1})^T\boldsymbol{\Gamma}^T\mathrm{d}s]\boldsymbol{P}^T. \tag{6.4.6}$$

Let $\boldsymbol{F}$ and $\boldsymbol{G}$ denote the matrix $\boldsymbol{P}^{-1}\boldsymbol{K}\boldsymbol{K}^T(\boldsymbol{P}^{-1})^T$ and $\boldsymbol{\Gamma}\boldsymbol{P}^{-1}\boldsymbol{K}\boldsymbol{K}^T(\boldsymbol{P}^{-1})^T\boldsymbol{\Gamma}^T$, respectively. Considering that $\boldsymbol{\Gamma}$ is a diagonal matrix, one deduces:

$$\boldsymbol{G}(i,j) = \boldsymbol{F}(i,j)\exp(\lambda_i t - \lambda_i s)\exp(\lambda_j t - \lambda_j s) = \boldsymbol{F}(i,j)\exp[(\lambda_i + \lambda_j)(t-s)] \tag{6.4.7}$$

where $\boldsymbol{F}(i, j)$ and $\boldsymbol{G}(i, j)$ denote the (i, j)th element of $\boldsymbol{F}$ and $\boldsymbol{G}$ in the ith row and jth column, respectively.

Let $\boldsymbol{H}$ denote the matrix $\int_{t_0}^{t} \boldsymbol{\Gamma}\boldsymbol{P}^{-1}\boldsymbol{K}\boldsymbol{K}^T(\boldsymbol{P}^{-1})^T\boldsymbol{\Gamma}^T\mathrm{d}s$. Furthermore, the integral of the element of $\boldsymbol{G}$ (6.4.7) can be obtained as:

$$\begin{aligned}\boldsymbol{H}(i,j) &= \int_{t_0}^{t} \boldsymbol{G}(i,j)\mathrm{d}s = \int_{t_0}^{t} \boldsymbol{F}(i,j)\exp[(\lambda_i + \lambda_j)(t-s)]\mathrm{d}s \\ &= \boldsymbol{F}(i,j)\{\exp[(\lambda_i + \lambda_j)(t-t_0)] - 1\}/(\lambda_i + \lambda_j)\end{aligned} \tag{6.4.8}$$

where $\boldsymbol{H}(i, j)$ denotes the (i, j)th element of $\boldsymbol{H}$.

Based on (6.4.8), $\boldsymbol{H}$ can be expressed as:

$$\boldsymbol{H} = \boldsymbol{F} \circ \boldsymbol{J} \tag{6.4.9}$$

where "$\circ$" denotes Hadamard product; and $\boldsymbol{J}$ denotes a matrix whose (i, j)th element is $\{\exp[(\lambda_i + \lambda_j)(t - t_0)] - 1\}/(\lambda_i + \lambda_j)$.

Substituting $\boldsymbol{H}$ (24) into $D[\Delta \boldsymbol{X}(t)]$ (6.4.6), one obtains:

$$D[\Delta \boldsymbol{X}(t)] = \boldsymbol{P}[\boldsymbol{F} \circ \boldsymbol{J}]\boldsymbol{P}^T = \boldsymbol{P}\{[\boldsymbol{P}^{-1}\boldsymbol{K}\boldsymbol{K}^T(\boldsymbol{P}^{-1})^T \circ \boldsymbol{J}]\}\boldsymbol{P}^T. \tag{6.4.10}$$

Equation (6.4.10) is the explicit formulation of the solution to LSDEs (6.4.2) system states covariance matrix, where there is no integral.

In the following, the stationary distribution of system states during the forced oscillation under stochastic disturbance will be deduced.

Because the system is small-signal stable, all the eigenvalues (i.e., λ_1, λ_2, …, λ_i, …) of system state matrix $\boldsymbol{A}$ have negative real parts. Thus, one obtains:

$$\lim_{t \to +\infty} \exp(\lambda_i t) = 0 \tag{6.4.11}$$

and

$$\lim_{t \to +\infty} \exp(\boldsymbol{A}t) = \boldsymbol{O} \tag{6.4.12}$$

By employing (6.4.12), the limit of system states mean in (6.4.3) can be deduced as:

$$\lim_{t \to +\infty} E[\Delta \boldsymbol{X}(t)] = \lim_{t \to +\infty} \exp[\boldsymbol{A}(t - t_0)]\Delta \boldsymbol{X}(t_0) = \boldsymbol{O}. \tag{6.4.13}$$

By using (6.4.11), the following formulation can be obtained.

$$\begin{aligned}\lim_{t \to +\infty} \boldsymbol{J}(i,j) &= \lim_{t \to \infty}\{\exp[(\lambda_i + \lambda_j)(t - t_0)] - 1\}/(\lambda_i + \lambda_j) \\ &= \{\lim_{t \to \infty} \exp[(\lambda_i + \lambda_j)(t - t_0)] - 1\}/(\lambda_i + \lambda_j) = -1/(\lambda_i + \lambda_j) = \boldsymbol{T}(i,j)\end{aligned} \tag{6.4.14}$$

where $\boldsymbol{J}(i, j)$ denotes the (i, j)th element of $\boldsymbol{J}$; and $\boldsymbol{T}$ denotes a matrix whose (i, j)th element $\boldsymbol{T}(i, j)$ is $-1/(\lambda_i + \lambda_j)$, which is finite due to the denominator's negative real part.

Substituting (6.4.14) into $\boldsymbol{D}[\Delta \boldsymbol{X}(t)]$ in (6.4.18), one can obtain the limit of system states covariance matrix as follows:

$$\lim_{t\to+\infty} D[\Delta\boldsymbol{X}(t)] = \boldsymbol{P}\{[\boldsymbol{P}^{-1}\boldsymbol{K}\boldsymbol{K}^T(\boldsymbol{P}^{-1})^T \circ \lim_{t\to\infty}\boldsymbol{J}]\}\boldsymbol{P}^T$$
$$= \boldsymbol{P}\{[\boldsymbol{P}^{-1}\boldsymbol{K}\boldsymbol{K}^T(\boldsymbol{P}^{-1})^T] \circ \boldsymbol{T}\}\boldsymbol{P}^T. \quad (6.4.15)$$

It can be seen that both the mean and covariance of system states have a convergence, which can be calculated by (6.4.13) and (6.4.15), respectively. By (6.4.15), the ith diagonal element in the $\lim_{t\to+\infty} D[\Delta\boldsymbol{X}(t)]$ represents the stationary variance of ith system state x_i in $\Delta\boldsymbol{X}(t)$; and the stationary mean of all system states equals to zero by (6.4.13). Based on that all system states of LSDEs follow Gaussian distributions, the stationary probability density function of ith system state x_i is deduced as follows:

$$f(x_i) = \exp[-x_i^2/\sigma^2]/(2\pi\sigma^2)^{0.5} \quad (6.4.16)$$

where σ^2 denotes the variance of ith system state x_i and equals to the ith diagonal element of $\lim_{t\to+\infty} D[\Delta\boldsymbol{X}(t)]$ (6.4.15).

Based on the three-sigma rule, "nearly all" values following a Gaussian distribution are taken to lie within three standard deviations, i.e., that it is useful to treat 99.7% probability as "near certainty". In other words, "nearly all" system states of LSDEs (6.4.10) fall within the corresponding three-sigma interval $[-3\sigma, 3\sigma]$, as shown in Fig. 6.2. Employing the stationary variance of the system states in (6.4.15), one can easily estimate the amplitude of system states during the forced oscillation under stochastic disturbance.

A modified SMIB power system is employed as the first simulation system. In this system, a wind farm connects with the infinite bus. The wind farm is equivalent to an asynchronous wind turbine generator. The stochastic disturbance is considered as the stochastic mechanical power inputs of the wind turbine generator. Then, the stochastic model of this power system can be expressed as LSDEs (6.4.10). There are three system states in this system, the real part of transient potential $\Delta E_r'$, the imaginary part of transient potential $\Delta E_m'$, and the slip of the induction generator Δs. Both the system state matrix $\boldsymbol{A}$ and the covariance matrix of stochastic disturbance $\boldsymbol{K}$ for (6.4.10) are shown as follows:

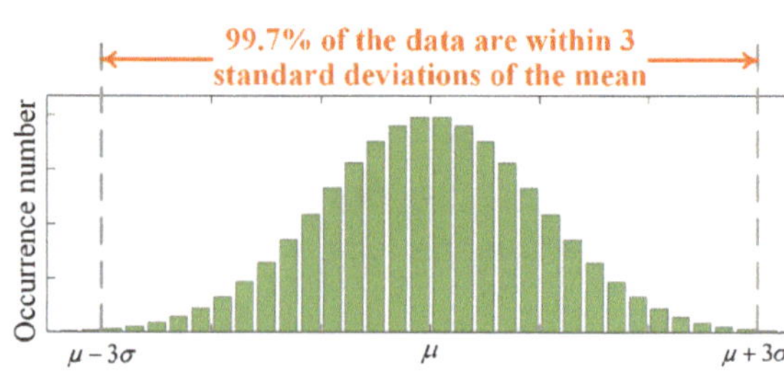

Fig. 6.2 Three-sigma rule for the Gaussian distribution

$$\boldsymbol{A} = \begin{bmatrix} -6.1678 & -2.2854 & -78.1942 \\ 2.2854 & -6.1678 & -267.69 \\ 0.0105 & 0.6632 & 0 \end{bmatrix} \text{ and } \boldsymbol{K} = \begin{bmatrix} 0 & 0 & 0 \\ 0 & 0 & 0 \\ 0 & 0 & 0.1642 \end{bmatrix} \quad (6.4.17)$$

where there is one nonzero element in the covariance matrix $\boldsymbol{K}$, because the stochastic disturbance describing the stochastic mechanical power inputs are added to the electromechanical equation.

The Heun's method is used to obtain trajectories of system states under stochastic disturbance. Figure 6.3a shows 20 trajectories of the slip of the induction generator Δs during the forced oscillation under stochastic disturbance, from which it is difficult to find the statistics directly. To illustrate the stochastic characteristics of system states during forced oscillation, Fig. 6.3b shows the probability density function of Δs during the forced oscillation. It can be seen that the probability density function of Δs converges to a certain distribution over time, which means the system state Δs has a stationary distribution. In other words, the oscillation behavior of a power system under stochastic disturbance is randomly "steady-state" after a certain time. By the stochastic analysis, the stationary distribution is theoretically explained and its probability density function analytically calculated by (6.4.16) directly.

The stationary distributions clearly reflect the characteristics of the forced oscillation under stochastic disturbance. Because the stationary distribution of a system state is a Gaussian distribution, the stationary variance and mean are of greatest interest for study. For the SMIB power system, the stationary variances and means of system states are calculated by Monte Carlo simulation and the proposed method, which are shown in Table 6.1. Clearly, the results of the analytic method are very close to Monte Carlo simulation results, which demonstrate that the proposed method is similarly accurate comparing with Monte Carlo simulation.

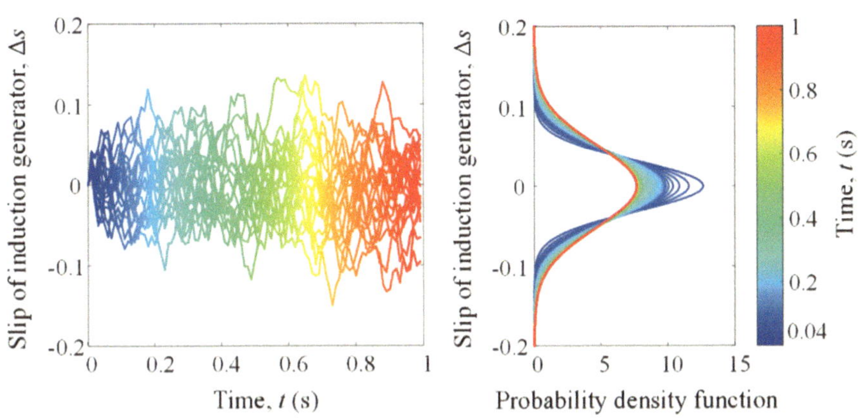

(a) 20 trajectories **(b) Probability density function**

Fig. 6.3 The slip of the induction generator Δs during the forced oscillation

Table 6.1 Comparisons of stationary variance and mean from Monte Carlo simulation and the proposed method

State variables	Stationary variances		Stationary mean	
	Monte Carlo simulation	The proposed method	Monte Carlo simulation	The proposed method
$\Delta E_r'$	5.5449×10^{-2}	5.5681×10^{-2}	-3.8349×10^{-3}	0
$\Delta E_m'$	9.4050×10^{-1}	9.3670×10^{-1}	-1.7646×10^{-3}	0
Δs	2.6660×10^{-3}	2.7137×10^{-3}	1.1155×10^{-3}	0

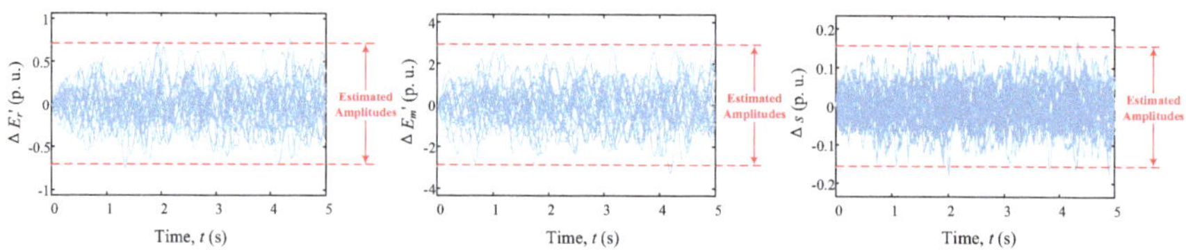

(a) ΔEr' (b) ΔEm' (c) Δs

Fig. 6.4 Amplitude estimation of system states under stochastic disturbance

Furthermore, the stationary standard deviation values of system states $\Delta E_r'$, $\Delta E_m'$, and Δs are calculated as 0.2360, 0.9678, and 0.0521, respectively. Based on the proposed amplitude estimation scheme, the amplitudes of $\Delta E_r'$, $\Delta E_m'$, and Δs are estimated as 0.7079, 2.9035, and 0.1563, respectively. In Fig. 6.4a–c, 20 trajectories and the corresponding estimated amplitudes of the three system states are shown by solid and dotted lines, respectively. It can be seen that most parts of trajectories are within the corresponding estimated amplitudes, which verifies the effectiveness of the proposed scheme.

To demonstrate the scalability of the proposed stochastic analytic method on multi-machine power systems under stochastic disturbance, a modified IEEE 14-bus case is adopted as shown in Fig. 6.5.

This modified case is included in MATLAB-based Power System Analysis Toolbox (PSAT). From PSAT, one also can find the detailed description of dynamic and static parameters for this case. In this modified IEEE 14-bus case, there is a wind farm (i.e., G1-DFIG) connecting to Bus 1, which is equivalent as a doubly fed induction generator with a variable speed wind turbine in PSAT. Also, there are four synchronous machines (i.e., G2, G3, G4, and G5) connecting to Bus 2, Bus 3, Bus 8, and Bus 6, which are modeled by fourth-order synchronous machine models. All loads are described by the constant impedance models. A simple eigenvalue analysis report from PSAT is given in Table 6.2. It can be seen that there are 48 system states and all eigenvalues have negative real parts in this modified IEEE 14-bus case.

The renewable energy generation from G1-DFIG is 90.8% of the total generation in the system, which means the system is highly penetrated by renewable energy. With the high penetration renewable energy in this system, the stochastic

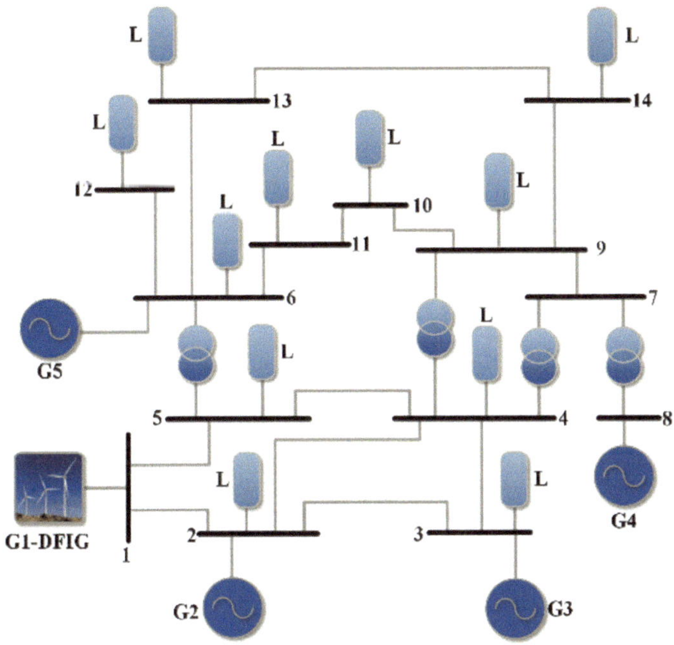

Fig. 6.5 A modified IEEE 14-bus case

Table 6.2 Eigenvalue analysis report of the modified IEEE 14-bus case

Name	Number
Dynamic order	48
Buses	14
Eigenvalues with positive real parts	0
Eigenvalues with negative real parts	48
Complex pairs of eigenvalues	9
Zero eigenvalues	0

disturbance becomes considerable. The stochastic disturbance assumed from the mechanical power input of wind turbine G1-DFIG, due to the stochastic nature of wind. Then the electromechanical equation of wind generator can be expressed as:

$$2H_m \mathrm{d}\omega = (P_m - P_e)\mathrm{d}t + \sigma \mathrm{d}B(t) \tag{6.4.18}$$

where H_m denotes rotor inertia;ω denotes angular frequency; P_m denotes mechanical power; P_e denotes electrical power; $\sigma \mathrm{d}B(t)$ denotes the stochastic disturbance; $B(t)$ denotes a standard Wiener process; and σ denotes the intensity of stochastic disturbance.

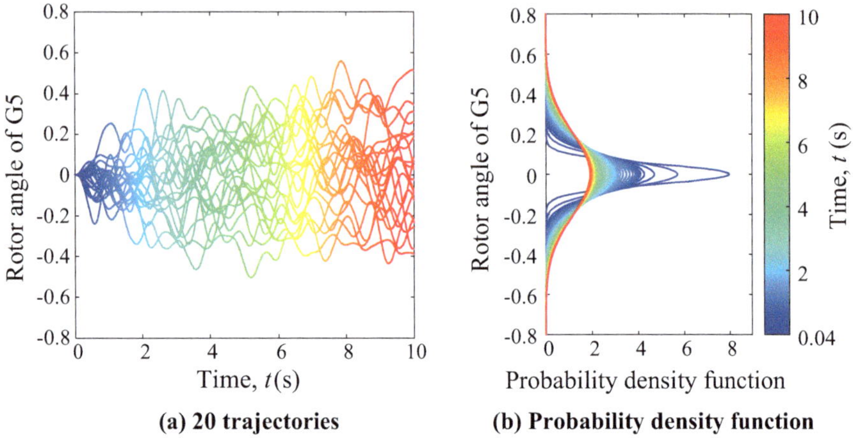

Fig. 6.6 The rotor angle of the generator G5 during the forced oscillation

In this simulation case, the intensity of the stochastic disturbance σ is set as 0.0333 P_m. After the consideration of stochastic disturbance, the system can be described by a set of SDAEs. However, no mature commercial software could simulate SDAEs directly. In this study, to implement the stochastic analysis of forced oscillation, PSAT is initially used to obtain the system state matrix $\boldsymbol{A}$, and then a separate MATLAB program is coded to simulate the LSDEs. PSAT is a user-friendly open-source MATLAB software program. One can export the system state matrix $\boldsymbol{A}$ from PSAT after the eigenvalue analysis is executed. Based on the exported system state matrix $\boldsymbol{A}$, the LSDEs model for this modified IEEE 14-bus case is built. Furthermore, the analysis of the forced oscillation under stochastic disturbance can be carried on with the LSDEs.

By using the Heun's method, 20 rotor angle trajectories of G5 are shown in Fig. 6.6a. Furthermore, the probability density function of rotor angle of G5 over time is shown in Fig. 6.6b, from which the phenomenon that the rotor angle has a stationary distribution is found. Actually, the same phenomenon is found for all the system states in this case, but only the rotor angle of G5 is illustrated here.

Let ω_1, ω_2, ω_3, ω_4, and ω_5 denote the rotor speed deviations of the five generators G1-DFIG, G2, G3, G4, and G5, respectively. By using Monte Carlo simulation and the proposed analytic method, the variances of ω_1, ω_2, ω_3, ω_4, and ω_5 are calculated, as shown in Table 6.3. It can be seen that the variances from Monte Carlo simulation get closer to the stationary variances from the proposed method, when the simulation time increases, time step decreases, and iteration increases.

The computation efficiency is also important in engineering application. For this modified IEEE 14-bus case with 48 system states, it takes 0.004 s to calculate

Table 6.3 Comparison of variances from Monte Carlo simulation and the proposed method

State variables	Variances from Monte Carlo simulation			Stationary variances from the proposed method
	Iteration 2500, simulation time 5 s, time step 0.002 s	Iteration 5000, simulation time 10 s, time step 0.001 s	Iteration 10,000, simulation time 20 s, time step 0.0005 s	
ω_1	8.1500×10^{-4}	1.1005×10^{-3}	1.1588×10^{-3}	1.1706×10^{-3}
ω_2	1.5248×10^{-6}	1.5865×10^{-6}	1.6223×10^{-6}	1.6158×10^{-6}
ω_3	4.5546×10^{-7}	4.5645×10^{-7}	4.5752×10^{-7}	4.5917×10^{-7}
ω_4	1.3697×10^{-6}	1.4018×10^{-6}	1.4469×10^{-6}	1.4342×10^{-6}
ω_5	1.1870×10^{-6}	1.2108×10^{-6}	1.2436×10^{-6}	1.2354×10^{-6}

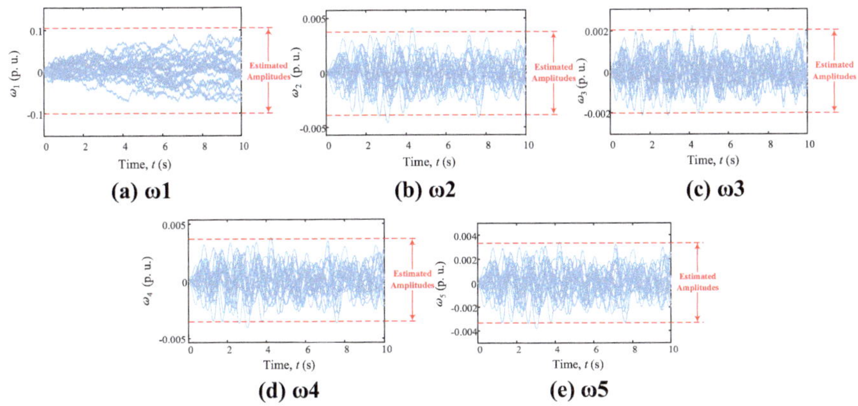

Fig. 6.7 Amplitude estimation of system states under stochastic disturbance

stationary variances by using the proposed method. However, it takes 301.009 s by Monte Carlo simulation with iteration 5000, simulation time 10 s, and time step 0.001 s, and the result accuracy by this Monte Carlo simulation is not very high, see Table 6.3. From the aspect of efficiency, the proposed analytic method is many orders of magnitude faster than Monte Carlo simulation.

With the calculated stationary variances of ω_1, ω_2, ω_3, ω_4, and ω_5, one can easily estimate the amplitudes of those system states as 0.1026, 0.0038, 0.0020, 0.0036 and 0.0033, respectively, by using the proposed amplitude estimation scheme. In Fig. 6.7a–e, 20 trajectories and the corresponding estimated amplitudes of the five system states are shown by solid and dotted lines, respectively. Clearly, most parts of system states' trajectories are within the corresponding estimated amplitudes, which verifies the effectiveness of the proposed scheme in a multi-machine power system under stochastic disturbance.

6.4.2 Intra-Region Probability of State Fluctuation in Power System

Based on the approximate assumption that the generator is represented by the classical second-order model, the load is represented by the constant impedance model and the stochastic disturbance is regarded as the Gaussian process, the model of the electromechanical transient process of the multi-machine power system under stochastic disturbance can be written as:

$$\begin{cases} \mathrm{d}\delta_i/\mathrm{d}t = \omega_N\omega_i \\ M_i\mathrm{d}\omega_i/\mathrm{d}t = P_i - \sum\limits_{j=1, j\neq i}^{n} E_iE_jB_{ij}\sin(\delta_i - \delta_j) - D_i\omega_i + \sigma_i W_i(t) \end{cases} \tag{6.4.19}$$

where $P_i = P_{mi} - G_{ii}E_i^2$; $W_i(t)$ and σ_i denote an independent standard Gaussian white noise and its intensity,respectively; $i = 1, 2, \ldots, n$. It should be pointed out that, except that M_i and t are named values, the remaining quantities are per-unit values. For the stochastic model (6.4.19), the energy function is chosen as:

$$\begin{aligned} H = & \frac{1}{2}\sum_{i=1}^{n} M_i\omega_N\omega_i^2 + \sum_{i=1}^{n} P_i(\delta_{is} - \delta_i) \\ & + \left[\sum_{i=1}^{n}\sum_{j=i+1}^{n} E_iE_jB_{ij}\cos(\delta_{is} - \delta_{js}) - \sum_{i=1}^{n}\sum_{j=i+1}^{n} E_iE_jB_{ij}\cos(\delta_i - \delta_j)\right] \end{aligned} \tag{6.4.20}$$

In a multi-machine power system, the boundary of a bound fluctuation region (or stable domain) is actually a hyperplane in a multi-dimension space. It is very difficult to accurately describe the boundary. As a result, the study on the relationship between the state of the system and the accurate domain becomes complex. According to the relationship between system states and energies, the research on complex high-dimension state vectors can be transformed into the one on simple one-dimension energy functions.

The system initial energy is set to $H(0)$ and the system energy at time t is defined as $H(t)$. If the system state $X(t)$ is within the bound fluctuation domain Ω_{B}, the energy $H(t)$ corresponding to the system state $X(t)$ is considered to be smaller than the energy H_{B} corresponding to the boundary of the bound fluctuation region. As shown in Fig. 6.8, the energy $H(t)$ of the system at time t has been greater than the energy H_{B} at the boundary of the bound fluctuation region, but less than the energy H_{cr} at the boundary of the stable domain. Clearly, the system state $X(t)$ has crossed the bound fluctuation region Ω_{B}.

The energy function of the power system is a function of system state δ_i and ω_i. If δ_i or ω_i are too large, the overall energy function H can be too large. Therefore, a certain small range of system energy can be indirectly used to describe whether the

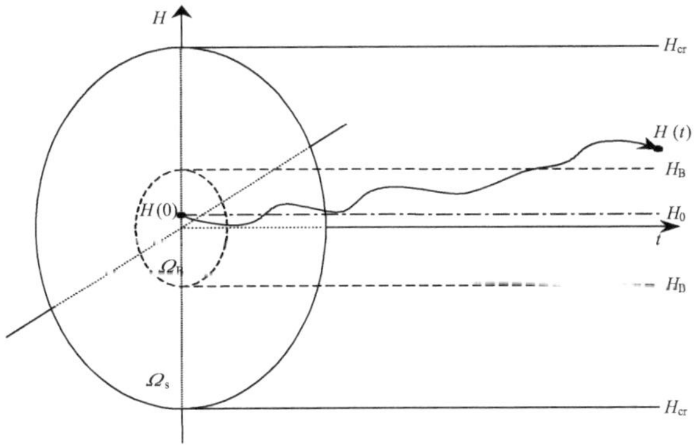

Fig. 6.8 Power system energy and energy boundary

system state is out of bounds or not, that is, whether the system state exceeds the bound fluctuation region or not.

In this study, it is defined that the energy corresponding to the boundary of the bound fluctuation region is $H_B = kH_{cr}(k < 1)$, and H_{cr} is the energy corresponding to the boundary of the system stability domain. Engineers can adjust the size of k (that is, adjust H_B) according to the needs of the operation. If the requirement for grid operation is high, the value of k can be correspondingly reduced (i.e., reducing the value of H_B) and vice versa. Furthermore, the intra-region probability of a bound fluctuation region can be transformed into a reliability function for $H(t)$, which can be expressed as

$$R(t|H_0) = P\{H(s) < H_B, s \in (0, t]|H(0) = H_0 < H_B\} \tag{6.4.21}$$

where H_0 denotes the energy of the initial state of the system; $R(t|H_0)$ denotes the probability that the system state is always within the bound fluctuation region at $(0, t]$.

$R(t|H_0)$ can be easily calculated using the Monte Carlo simulation, but the computation is intensive for multi-machine power systems. In order to solve $R(t|H_0)$ quickly, an analytical algorithm based on the quasi Hamiltonian SAM is proposed in the research to solve the intra-region probability $R(t|H_0)$ analytically. According to the quasi Hamiltonian system theory: it can be regarded as the quasi Hamiltonian system as long as, during one cycle period of vibration, the energy input into the system by the stochastic disturbance and the energy consumed by the damping are smaller than the total energy of the system itself. In a real power system, the stochastic disturbance and the product of the damping and rotation speed are smaller than the mechanical power and electromagnetic power of the system, so the multi-machine power system under the stochastic disturbance can be regarded as the quasi Hamiltonian system. The stochastic model (6.4.19) can be transformed into a quasi Hamiltonian model as follows:

$$\begin{cases} d\delta_i = \frac{1}{M_i}\frac{\partial H}{\partial \omega_i}dt \\ d\omega_i = \left(-\frac{1}{M_i}\frac{\partial H}{\partial \delta_i} - \frac{D_i}{M_i^2\omega_N}\frac{\partial H}{\partial \omega_i}\right)dt + \frac{\sigma_i}{M_i}dB_i(t) \end{cases} \quad i = 1, 2, \ldots, n \tag{6.4.22}$$

Based on the quasi Hamiltonian system SAM, the system energy H (Hamiltonian function) weakly converges to a one-dimension diffusion process:

$$dH = \bar{m}(H)dt + \bar{\sigma}(H)dB(t) \tag{6.4.23}$$

where $\bar{m}(H)$ and $\bar{\sigma}(H)$ that are the drift coefficient and the diffusion coefficient for the diffusion process, respectively, which can be obtained as follows:

$$\begin{cases} \bar{m}(H) = \left(\sum_{i=1}^{n}\frac{\omega_N\sigma_i^2}{2M_i}\right) - \left[\sum_{i=1}^{n}\frac{2D_i}{M_i(2n-1)}\right]H \\ \bar{\sigma}^2(H) = \left[\sum_{i=1}^{n}\frac{2\omega_N\sigma_i^2}{M_i(2n-1)}\right]H \end{cases} \tag{6.4.24}$$

In the above equation, there is only a simple algebraic operation, without complicated integral calculation. Therefore, the computational efficiency is greatly increased, which also provides the basis for the large-scale power system to apply the SAM.

For Eq. (6.4.23), by using the Kolmogorov's backward equation, the analytical equation for $R(t|H_0)$ can be obtained as:

$$\frac{\partial R(t|H_0)}{\partial t} = \bar{m}(H_0)\frac{\partial R(t|H_0)}{\partial H_0} + \frac{\bar{\sigma}(H_0)^2}{2}\frac{\partial^2 R(t|H_0)}{\partial H_0^2} \tag{6.4.25}$$

In such equation, $\bar{m}(H_0)$ and $\bar{\sigma}(H_0)$ can be obtained by calculation of Eq. (6.4.24), and change with H_0. Equation (6.4.25) is a parabolic partial differential equation, whose completed analytical result is difficult to deduce directly and usually solved via numerical solutions such as the Crank-Nicolson formula of the finite difference method. Both the initial value of the equation and its boundary need to be clear before solving. The initial value of Eq. (6.4.25) and the boundary conditions are as follows:

$$\begin{cases} \text{Initial value: when } H_0 < H_B \; R(0|H_0) = 1 \\ \text{Upper boundary: } R(t|H_0) = 0 \\ \text{Lower boundary: when } H_0 = 0 \; \frac{\partial R}{\partial t} = \bar{m}(0)\frac{\partial R}{\partial H_0} \end{cases} \tag{6.4.26}$$

A four-machine two-area system is adopted as the simulation example, in which the system load is regarded as a constant impedance load, and a second-order rotor motion equation is applied for the generator. By the network simplification, nodes of the generators are maintained, and stochastic power fluctuations are transfered to each internal node of generators. An Itô equation of four-machine power system subjected to stochastic disturbances can be obtained. The intensities of stochastic

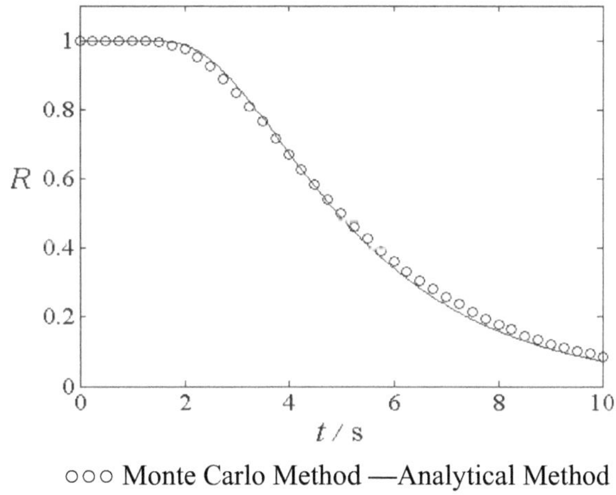

Fig. 6.9 Comparison of analytical method with Monte Carlo simulation

disturbances are set to $\sigma_1 = 0.03$, $\sigma_2 = 0.04$, $\sigma_3 = 0.05$ and $\sigma_4 = 0.06$. Comparing the Monte Carlo simulation with a analytical solution, the results are shown in Fig. 6.9. It can be seen that the analytical solution is well fitted to the Monte Carlo solution. What's more, the Monte Carlo simulation's computation (taken for 5000 times) time is about 246.35 s, while the analytical method only needs about 0.33 s.

6.4.3 Intra-Region Probability of Frequency Fluctuation in Power System

Frequency is one of the most important indexes of the power system. There are many reasons causing system dynamic frequency fluctuations, but the stochastic power fluctuation (stochastic disturbance) is the most direct and main factor. In a traditional power system, the load change can cause stochastic disturbance, but its fluctuation amplitude is generally small. In recent years, with the integration of renewable energy sources and electric vehicles, uncontrollable stochastic disturbances gradually become obvious, and their impacts on the dynamic security of system frequencies attract attentions.

By using the per-unit value, the frequency can be considered equal to the rotation speed. The dynamic frequency of a single generator is its generator speed. The dynamic frequency of the entire power system can be described by the frequency of the system's equivalent inertia center. It refers to the rotating speed produced by the sum of the unbalanced power of the system acting upon the system's equivalent inertia center. The system dynamic frequency increment Δf can be expressed by the following equation:

$$\Delta f = \omega = \left(\sum_{i=1}^{m} M_i\omega_i\right) \Big/ \left(\sum_{i=1}^{m} M_i\right) \tag{6.4.27}$$

where ω denotes an inertial center speed deviation; ω_i and M_i denote the ith generator's speed deviation and inertia time constant, respectively; m denotes the number of generators within the system (or region). It should be pointed out that, except that M_i is a named value, Δf, ω and ω_i are per-unit values. For the convenience of narration, ω should be employed to indicate Δf hereinafter.

Due to the stochastic disturbance, the dynamic frequency of the system is stochastic at a certain time and no longer a deterministic value. Therefore, the system dynamic frequency within the security permission range is a stochastic event, which also needs to be described by the intra-region probability. The probability that the system dynamic frequency is always within the security region is called the dynamic frequency intra-region probability $R(t|\omega_0)$, and $R(t|\omega_0)$ can be expressed as the probability where the initial value ω_0 of dynamic frequency ω lies within the range of $\Omega = [\omega_{\min}, \omega_{\max}]$, and the dynamic frequency remains within Ω at time $(0, t]$, i.e.,

$$R(t|\omega_0) = P\{\omega(s) \in \Omega, s \in (0, t] | \omega(0) = \omega_0 \in \Omega\} \tag{6.4.28}$$

where $\omega(s)$ represents the system dynamic frequency at time s.

After some approximate assumptions, the model of the power system with m generator nodes and n load nodes under stochastic disturbances is as follows:

$$\begin{cases} \dfrac{d\delta_i}{d_t} = \omega_N\omega_i & i = 1, 2, \ldots, m \\ M_i\dfrac{d\omega_i}{d_t} = P_{mi} - \sum_{j=1, j\neq i}^{m+n} P_{\max ij}S_{\delta\, ij} - D_i\omega_i + \sigma_i W_i(t) & i = 1, 2, \ldots, m \\ 0 = P_{li} - \sum\limits_{j=1, j\neq i}^{m+n} P_{\max ij}S_{\delta\, ij} + \sigma_i W_i(t) & i = m+1, m+2, \ldots, m+n \end{cases} \tag{6.4.29}$$

where $P_{\max ij} = E_iE_jB_{ij}$; $S_{\delta ij} = \sin(\delta_i - \delta_j)$; $\sigma_i W_i(t)$ denotes the stochastic disturbance; $W_i(t)$ and σ_i denote the standard Gaussian white noise and its intensity, respectively.

According to the result of the model (6.4.29), by substituting it into (6.4.27), we can obtain the system dynamic frequency ω. However, the model (6.4.29) is a high-order nonlinear Itô-type stochastic differential algebraic equation, which is extremely difficult to solve analytically and directly by existing research methods. The average system frequency model can be obtained through simplification by the full state model (6.4.29), that is, the equations of all nodes in the entire network are equivalently aggregated into a single-machine model. By adding all power equations of different nodes of the model (6.4.29), it gives:

$$\left(\sum_{i=1}^{m} M_i\right)\mathrm{d}\omega = -\sum_{i=1}^{m} (D_i\omega_i)\mathrm{d}t + \sum_{i=1}^{m+n} [\sigma_i \mathrm{d}B_i(t)] \quad (6.4.30)$$

where $B_i(t)$ represents a Standard Wiener process, the derivative of $B_i(t)$ is denoted as $B_i(t)/\mathrm{d}t = W_i(t)$.

Although the model (6.4.30) is more simplified than the model (6.4.29), (6.4.30) cannot be analyzed analytically because ω_i of each generator node is still unknown. Assume that the generator has uniform damping, that is, each generator node sat isfies the condition of $D_i/M_i = k$. When $D_i = kM_i$ is substituted into Eq. (6.4.30), one can obtain:

$$\mathrm{d}\omega = -k\omega \mathrm{d}t + \sigma \mathrm{d}B(t) \quad (6.4.31)$$

where the size of σ depends on the correlation of M_i, σ_i and $W_i(t)$. For convenience narration, the stochastic disturbance $\sigma_i W_i(t)$ are assumed irrelevant to each other, that is, $\sigma = \sqrt{\sum_{i=1}^{m+n} \sigma_i^2} \Big/ \sum_{i=1}^{m} M_i$.

Although the uniform damping assumption is regularly used in power system research, the uniform damping assumption in the actual power system is relatively harsh. For a system with small damping inhomogeneity, the generator damping approximation is assumed to be $\tilde{D}_i = M_i \sum_{i=1}^{m} (D_i/M_i)/m$. That is, k_i (formally D_i/M_i) of each generator node is uniformly set to mean $\bar{k} = \sum_{i=1}^{m} k_i/m$. Furthermore, the original system approximates a uniform damping system, and the model (6.4.31) can still be used, in which $k = \bar{k} = \sum_{i=1}^{m} (D_i/M_i)/m$.

Model (6.4.31) is actually a first-order diffusion process, and $-k\omega$ is its drift coefficient, σ is its diffusion coefficient. By using the Kolmogorov's backward equation, the intra-region probability can be calculated. Then the reliability function $R(t|\omega_0)$ of Eq. (6.4.31) is:

$$\frac{\partial R(t|\omega_0)}{\partial t} = -k\omega_0 \frac{\partial R(t|\omega_0)}{\partial \omega_0} + \frac{\sigma^2}{2}\frac{\partial^2 R(t|\omega_0)}{\partial \omega_0^2} \quad (6.4.32)$$

where ∂ represents the partial derivative calculation.

The simulation example is a modified 4-machine and 2-area system. After the network is simplified (retaining the generator and load nodes), a system with 6 nodes is obtained. Stochastic disturbances of generator nodes are regarded as stochastic power fluctuations from the power generation part, and their intensities are σ_1, σ_2, σ_3 and σ_4 respectively. Stochastic disturbances of load nodes are regarded as stochastic power fluctuations from the load part, and the intensities are respectively σ_5 and σ_6 respectively. Finally, a stochastic differential algebraic equation with a 4-machine and 2-area system subjected to stochastic disturbances is obtained, as shown in (6.4.29).

Comparing the Monte Carlo simulation with the analytical method, the results are shown in Fig. 6.10. It can be seen that the Monte Carlo method and the analytic

Fig. 6.10 Comparison of analytical solution with Monte Carlo solution

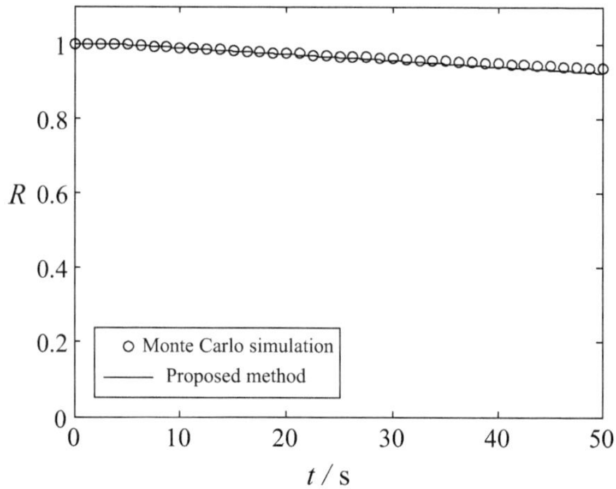

Fig. 6.11 Comparison of the results of non-uniform damping analysis method with the Monte Carlo simulation

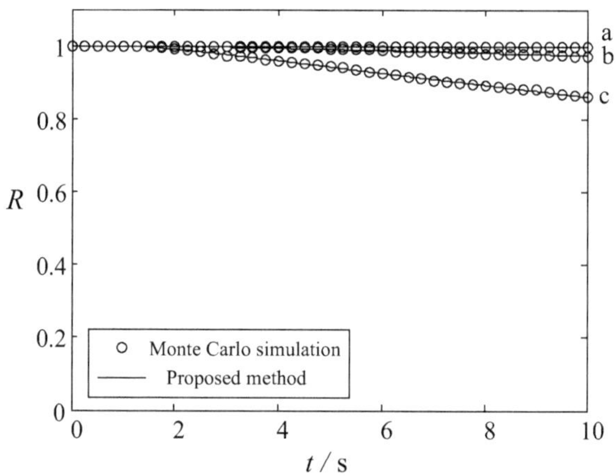

a - small stochastic disturbance; b - moderate stochastic disturbance; c - large stochastic disturbance.

method fit very well. What's more, the Monte Carlo simulation (5000 iterations, 0.002 s of simulation step) costs about 292.76 s, while the analytical method only needs about 0.015 s.

In the analytical method, the damping approximation is set to $\tilde{D}_i = M_i \sum_{i=1}^{m} (D_i/M_i)/m$. Taking the 4-machine and 2-zone system as an example, when the difference between k_i and $\bar{k}$ is no more than 20% (see Table 6.4 for parameter settings), the Monte Carlo simulation and analytical method under different stochastic disturbances are shown in Fig. 6.11. It should be noted that the analytic method adopts approximate damping $\tilde{D}_i$, while the Monte Carlo simulation still uses the original system damping D_i. It can be seen that the analytical solution and

Table 6.4 Comparison of the efficiency of Monte Carlo simulation with the analytical method

		Machine 2	Machine 3	Machine 4
Original damping D_i	7.8	7.8	4.94	4.94
Approximate damping used in the analytical method $\tilde{D}_i$	6.5	6.5	6.175	6.175
M_i	13	13	12.35	12.35
$k_i = D_i/M_i$	0.6	0.6	0.4	0.4
$\bar{k}$	0.5			

the Monte Carlo solution still fit well when the system does not satisfy the uniform damping assumption.

6.5 Dynamic Security of Power System under Large Stochastic Disturbance

The stability analysis of power systems under large disturbances is a widely recognized challenge in theory and engineering. When various stochastic factors are considered in the power system, the stability analysis becomes more complicated. In order to simplify the analysis, the quasi Hamilton system SAM and EEAC equivalent simplification methods are used in this section for the dynamic security analysis of power systems under large stochastic disturbances.

6.5.1 Stability Probability of Lossless System under Large Stochastic Disturbance

It is known that contingencies can greatly impact power system transient stability, but power system transient stability also faces new problems with increasing stochastic disturbance. It was proved that even if a dynamic system can tolerate one severe fault at its equilibrium point, cumulative effects of stochastic disturbance can cause its state trajectories to leave any bound domain after a sufficiently long time. For instance, a single-machine infinite-bus system is subjected to a three-phase fault on the transmission line for 150 ms, and its rotor angle trajectory is stable as shown in Fig. 6.12. However, when the same system is subjected to stochastic disturbance, one of 10 trajectories exhibits unexpected behavior and becomes unstable as shown in Fig. 6.13. Therefore, transient stability assessment methods considering only system faults are incomplete, which have difficulties in handling the stochastic disturbance.

Although transient stability assessment of power systems under stochastic disturbance has been considered and partly solved in the last decades, there are still

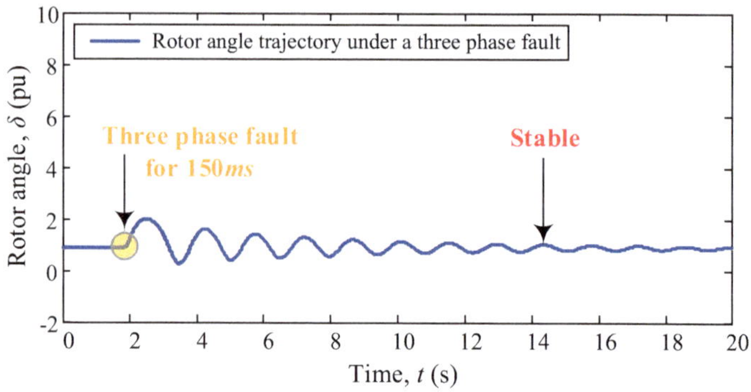

Fig. 6.12 Rotor angle trajectory of a power system under a three-phase fault

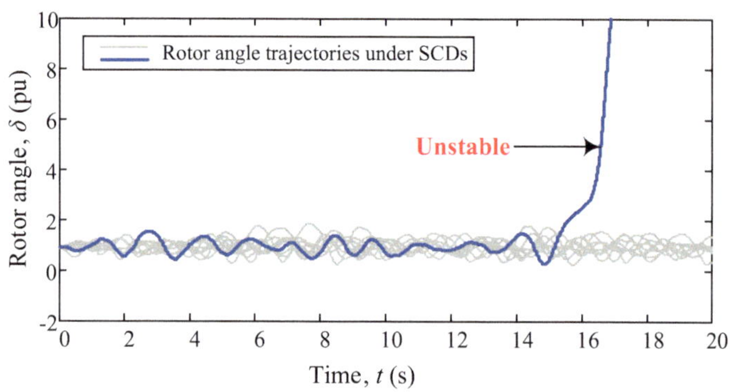

Fig. 6.13 Rotor angle trajectories of a power system under stochastic disturbance

some deficiencies in existing methods: (1) the majority of assessment methods are based on Monte Carlo simulation, which suffers low computational efficiency and unclear impact mechanism; (2) the application of analytical assessment methods in multi-machine power systems encounters difficulties, due to their complexity; (3) the probability measure of transient stability under stochastic disturbance has not been proposed. To address the deficiencies listed above, this study proposes a novel analytical transient stability assessment method for multi-machine power systems under stochastic disturbance.

In this study, the classical model has been adopted initially. Then the stochastic model of multi-machine power systems under stochastic disturbance can be expressed as a set of Itô stochastic differential equations:

$$\begin{cases} \frac{d\delta_i}{d_t} = \omega_N \omega_i \\ M_i \frac{d\omega_i}{d_t} = P_{mi} - G_{ii}E_i^2 - \sum\limits_{j=1, j\neq i}^{n} P_{\max ij} \sin(\delta_i - \delta_j) - D_i\omega_i + \sigma_i W_i(t) \end{cases} i = 1, 2, \ldots, n \tag{6.5.1}$$

where $\sigma_i W_i(t)$ are stochastic disturbance; $W_i(t)$ is standard Gaussian white noises; σ_i are disturbance intensities; $P_{\max ij} = E_i E_j B_{ij}$; δ_i, ω_i, M_i, P_{mi}, E_i and D_i are rotor angle, rotor speed deviation, inertia coefficient, mechanical power, internal voltage and damping coefficient of ith generator, respectively; G_{ii} is the ith diagonal element of conductance matrix; B_{ij} is the element in the ith row and jth column of susceptance matrix; and ω_N is the synchronous machine speed.

Because the network and load are merged, stochastic disturbance $\sigma_i W_i(t)$ is added to machine equations. Because the stochastic disturbance is from the mechanical power P_{mi} and the quadratic term of self-conductance $G_{ii}E_i^2$, they denote the stochastic unbalanced active power between the generation and load.

In the conventional transient stability assessment method, the system is either stable or unstable at a certain time. However, when the stochastic disturbance is introduced to a power system, one cannot answer whether the system is deterministically stable or unstable in the following time because the transient stability becomes a random event. The probability of states being within the stability region represents stability probability. From the view of operational decision-making, stability probability is a proper measure to evaluate the transient stability under stochastic disturbance. It is reasonable that the system with higher stability probability is more probabilistically stable. Stability probability can be expressed by the following form:

$$P(t|\boldsymbol{X}_0) = P\{\boldsymbol{X}(\tau) \in \Omega, \tau \in (0, t] | \boldsymbol{X}_0 \in \Omega\} \tag{6.5.2}$$

where $P(t|\boldsymbol{X}_0)$ is the probability of states being in the stability region Ω, within the time interval [0, t), given that the initial states $\boldsymbol{X}(0)$ are within Ω.

Energy function methods are well-suited to evaluate transient stability, by which high-dimension states are transformed into a single index of system energy. If system energy is lower than the critical energy, system states stay in the stability region. The critical energy corresponds to stability boundary, and its accurate computation is nontrivial. The corresponding energy function for (6.5.1) is as follows:

$$\begin{aligned} H = & \frac{1}{2}\sum_{i=1}^{n} M_i \omega_N \omega_i^2 + \sum_{i=1}^{n} P_i(\delta_{is} - \delta_i) \\ & + \left[\sum_{i=1}^{n}\sum_{j=i+1}^{n} E_i E_j B_{ij} \cos(\delta_{is} - \delta_{js}) - \sum_{i=1}^{n}\sum_{j=i+1}^{n} E_i E_j B_{ij} \cos(\delta_i - \delta_j)\right] \end{aligned} \tag{6.5.3}$$

where δ_{is} and δ_{js} denote the equilibrium rotor angle of the ith and jth generators, respectively.

Accordingly, the stability probability can be calculated by the following form:

$$P(t|H_0) = P\{H(\tau) < H_{\rm cr}, \tau \in (0,t] | H(0) = H_0 < H_{\rm cr}\} \tag{6.5.4}$$

where $P(t|H_0)$ is the conditional probability of $H(t)$ being lower than the critical energy $H_{\rm cr}$ within the time interval [0, t), given that the initial value H_0 is lower than $H_{\rm cr}$.

To apply SAM, the stochastic model should be expressed in the quasi-Hamiltonian form. Stochastic model in quasi-Hamiltonian form for (6.5.1) is as follows:

$$\begin{cases} {\rm d}\delta_i = \frac{1}{M_i}\frac{\partial H}{\partial \omega_i}{\rm d}t \\ {\rm d}\omega_i = \left(-\frac{1}{M_i}\frac{\partial H}{\partial \delta_i} - \frac{D_i}{M_i^2\omega_N}\frac{\partial H}{\partial \omega_i}\right){\rm d}t + \frac{\sigma_i}{M_i}{\rm d}B_i(t) \end{cases}, \quad i = 1,2,\ldots,n \tag{6.5.5}$$

where ∂ is the partial differential operator; $B_i(t)$ are Wiener processes, and ${\rm d}B_i(t)/{\rm d}t = W_i(t)$.

Based on Itô formula, Itô equation of system energy H can be expressed as follows:

$$\begin{aligned} {\rm d}H &= \sum_{i-1}^{n}\left\{\frac{\partial H}{\partial \omega_i}\left[\left(-\frac{1}{M_i}\frac{\partial H}{\partial \delta_i} - \frac{D_i}{M_i^2\omega_N}\frac{\partial H}{\partial \omega_i}\right){\rm d}t + \frac{\sigma_i}{M_i}{\rm d}B_i(t)\right] + \frac{1}{M_i}\frac{\partial H}{\partial \delta_i}\frac{\partial H}{\partial \omega_i}{\rm d}t + \frac{1}{2}\frac{\sigma_i^2}{M_i^2}\frac{\partial^2 H}{\partial \omega_i^2}{\rm d}t\right\} \\ &= \sum_{i=1}^{n}\left[-D_i\omega_N\omega_i^2 + \frac{\sigma_i^2\omega_N}{2M_i}\right]{\rm d}t + \sum_{i=1}^{n}\sigma_i\omega_N\omega_i{\rm d}B_i(t). \end{aligned} \tag{6.5.6}$$

Based on a lemma of Khasminskii, system energy H weakly converges to a first-order diffusion process, as follows:

$${\rm d}H = m^*(H){\rm d}t + \sigma^*(H){\rm d}B(t). \tag{6.5.7}$$

The coefficients $m^*(H)$ and $\sigma^*(H)$ can be obtained by time-averaging, where

$$\begin{cases} m^*(H) = \lim\limits_{T\to\infty}\frac{1}{T}\int\limits_0^T\sum\limits_{i=1}^{n}\left(-D_i\omega_N\omega_i^2 + \frac{\sigma_i^2\omega_N}{2M_i}\right){\rm d}t \\ \sigma^{*2}(H) = \lim\limits_{T\to\infty}\frac{1}{T}\int\limits_0^T\sum\limits_{i=1}^{n}(\sigma_i\omega_N\omega_i)^2{\rm d}t \end{cases}. \tag{6.5.8}$$

It is difficult to calculate time-averaging results ($m^*(H)$ and $\sigma^{*2}(H)$) from (6.5.8) directly because system states ω_i are unknown at a given time. Due to the ergodicity of Hamiltonian system, the following three hypotheses of ergodicity can be adopted to calculate the time-averaging results: (1) time-averaging results are equal to corresponding space-averaging results in the space $\Omega = \{(\Delta\delta_1, \ldots, \Delta\delta_{n-1}, \omega_1, \ldots, \omega_n) | H(\Delta\delta_1, \ldots, \Delta\delta_{n-1}, \omega_1, \ldots, \omega_n) = H\}$, where $\Delta\delta_i = \delta_i - \delta_n$ (the nth generator is regarded as a reference node); (2) system states can arrive everywhere in Ω; furthermore, (3) for any given energy level H, the conditional probabilities of different system state in Ω are equal. Let $\mathrm{d}s/S$ denote the probability of states being in a certain small piece, where S is the surface area of whole hypersphere Ω and $\mathrm{d}s$ is the surface area of a certain small piece of the hypersphere Ω. Then, space-averaging results, which are equal to time-averaging results, can be deduced as follows:

$$\begin{cases} \bar{m}(H) = \frac{1}{S}\int\limits_{\Omega}\sum\limits_{i=1}^{n}\left(-D_i\omega_N\omega_i^2 + \frac{\sigma_i^2\omega_N}{2M_i}\right)\mathrm{d}s \\ \bar{\sigma}^2(H) = \frac{1}{S}\int\limits_{\Omega}\sum\limits_{i=1}^{n}(\sigma_i\omega_N\omega_i)^2\mathrm{d}s \\ S = \int\limits_{\Omega} 1\mathrm{d}s \end{cases} \tag{6.5.9}$$

where $\int$ denotes the surface integral operator; Ω denotes the integral domain.

Accordingly, the first-order diffusion process of system energy becomes:

$$\mathrm{d}H = \bar{m}(H)\mathrm{d}t + \bar{\sigma}(H)\mathrm{d}B(t) \tag{6.5.10}$$

where $\bar{m}(H)$ and $\bar{\sigma}(H)$ are the drift coefficient and diffusion coefficient, respectively.

Based on many simulations for stochastic averaging results of (6.5.9), it is found that the relation between coefficients (i.e., $\bar{m}(H)$ and $\bar{\sigma}(H)$) and energy H is approximately linear, even though the original system is nonlinear. For instance, in the Kundur's 4-machine 2-area system, the stochastic averaging results are shown in Fig. 6.14. Clearly, the linear relation exists even when the system energy is close to critical energy.

The stochastic disturbance is usually smaller than system faults, which is one of the reasons that linear relation exists. It is known that a power system shows nonlinearity when a severe system fault occurs. The stability analysis under a severe system fault is a nonlinear problem because the system fault is large and instant. However, a novel instability mechanism works in the system under stochastic

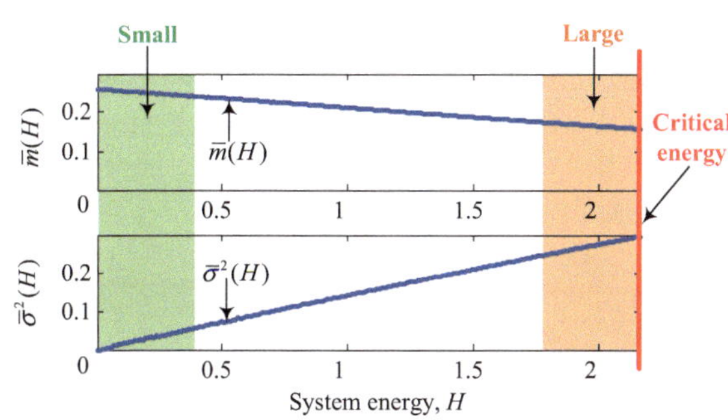

Fig. 6.14 Coefficients changing with system energy H

disturbance, which describes that the random effects of small stochastic disturbance "can cause the trajectories of the system to leave any bound domain with probability one", even the stable domain. According to model (6.5.1), the linearized stochastic model can be derived as follows:

$$\begin{cases} \frac{d\Delta\delta_i}{d_t} = \omega_N \Delta\omega_i & i = 1, 2, \ldots, n-1 \\ M_i \frac{d\Delta\omega_i}{d_t} = -\sum\limits_{j=1, j\neq i}^{n} C_{ij}(\Delta\delta_i - \Delta\delta_j) - D_i\Delta\omega_i + \sigma_i W_i(t) & i = 1, 2, \ldots, n \end{cases} \tag{6.5.11}$$

where $\Delta\omega_i = \omega_i$; $\Delta\delta_i = \delta_i - \delta_n$ and $\Delta\delta_n = 0$; δ_n is the rotor angle of nth generator, which is set as the reference node; and $C_{ij} = E_iE_jB_{ij}\cos(\delta_{is} - \delta_{js})$.

The linearized energy function is as follows:

$$H = \frac{1}{2}\sum_{i=1}^{n} M_i\omega_N\Delta\omega_i^2 + \frac{1}{2}\sum_{i=1}^{n}\sum_{j=i+1}^{n} C_{ij}(\Delta\delta_i - \Delta\delta_j)^2. \tag{6.5.12}$$

System energy (6.5.12) can be expressed as follows:

$$H = \boldsymbol{X}^T\boldsymbol{Q}\boldsymbol{X} \tag{6.5.13}$$

where $\boldsymbol{X} = (\Delta\delta_1, \ldots, \Delta\delta_{n-1}, \Delta\omega_1, \ldots, \Delta\omega_n)^T$; $\boldsymbol{Q}$ is a $(2n-1)$th-order positive definite matrix with the form,

$$\boldsymbol{Q} = \left(\begin{array}{c|c} \boldsymbol{Q}_{11} & \boldsymbol{O} \\ \boldsymbol{O} & \boldsymbol{Q}_{22} \end{array}\right) \tag{6.5.14}$$

where $\boldsymbol{Q}_{11}$ is an $(n-1)$th-order nonsingular symmetric matrix; $\boldsymbol{Q}_{22}$ is a nth-order diagonal matrix with the form $\boldsymbol{Q}_{22} = diag(M_1\omega_N/2, \ldots, M_n\omega_N/2)$; and $\boldsymbol{O}$ denotes a null matrix.

Differentially (6.5.13), one obtains:

$$\begin{cases} \partial H/\partial \boldsymbol{X} = 2\boldsymbol{Q}\boldsymbol{X} \\ \partial^2 H/\partial \boldsymbol{X}^2 = 2\boldsymbol{Q} \end{cases} \tag{6.5.15}$$

where $\partial H/\partial \boldsymbol{X}$ is a $(2n-1)$th-order column vector, and $\partial H/\partial \boldsymbol{X} = (\partial H/\partial\Delta\delta_1, \ldots, \partial H/\partial\Delta\delta_{n-1}, \partial H/\partial\Delta\omega_1, \ldots, \partial H/\partial\Delta\omega_n)^T$; $\partial^2 H/\partial \boldsymbol{X}^2$ is a $(2n-1)$ th-order square matrix, in which the element in ith row and jth column is $\partial^2 H/\partial x_i\partial x_j$. x_i and x_j denote the ith and jth element of $\boldsymbol{X}$, respectively.

According to (6.5.15), the linearized stochastic model in quasi-Hamiltonian form is as follows:

$$\mathrm{d}\boldsymbol{X} = \boldsymbol{M}\partial H/\partial \boldsymbol{X}\mathrm{d}t + \boldsymbol{D}\partial H/\partial \boldsymbol{X}\mathrm{d}t + \boldsymbol{G}\mathrm{d}\boldsymbol{B}(t) \tag{6.5.16}$$

where $\mathrm{d}\boldsymbol{X} = (\mathrm{d}\Delta\delta_1, \ldots, \mathrm{d}\Delta\delta_{n-1}, \mathrm{d}\Delta\omega_1, \ldots, \mathrm{d}\Delta\omega_n)^T$; $\mathrm{d}\boldsymbol{B}(t) = [\mathrm{d}B_1(t), \ldots, \mathrm{d}B_i(t), \ldots, \mathrm{d}B_{2n-1}(t)]^T$ and $\mathrm{d}B_i(t)$ is independent under different i; $\boldsymbol{M}$, $\boldsymbol{D}$, and $\boldsymbol{G}$ are $(2n - 1)$th-order square matrices; $\boldsymbol{M}$ describes the characteristic of system conservation and $\boldsymbol{M} + \boldsymbol{M}^T = \boldsymbol{O}$; $\boldsymbol{D}$ describes the characteristic of system dissipation and is a $(2n - 1)$th-order diagonal matrix $diag(0, \ldots, 0, -D_1/\omega_N M_1^2, \ldots, -D_n/\omega_N M_n^2)$; $\boldsymbol{G}$ denotes the intensity of stochastic disturbance and is a $(2n - 1)$th-order diagonal matrix $diag(0, \ldots, 0, \sigma_1/M_1, \ldots, \sigma_n/M_n)$.

Combining (6.5.15) and (6.5.16), one has

$$\mathrm{d}\boldsymbol{X} = 2\boldsymbol{MQX}\mathrm{d}t + 2\boldsymbol{DQX}\mathrm{d}t + \boldsymbol{G}\mathrm{d}\boldsymbol{B}(t). \tag{6.5.17}$$

Given that $\boldsymbol{Q}$ in (6.5.14) is a positive definite matrix, the linearized energy function (6.5.13) can be transformed into the following standard quadratic form:

$$H = \boldsymbol{Y}^T\boldsymbol{Y} = y_1^2 + \cdots y_i^2 \cdots + y_{2n-1}^2 \tag{6.5.18}$$

where $\boldsymbol{Y} = (y_1, \ldots, y_i, \ldots, y_{2n-1})^T$, $\boldsymbol{Y} = \boldsymbol{CX}$, $\boldsymbol{Q} = \boldsymbol{C}^T\boldsymbol{C}$, and $\boldsymbol{C}$ is a non-singular matrix with $(2n - 1)$th-order. Due to the special form of $\boldsymbol{Q}$, $\boldsymbol{C}$ has the form of

$$\boldsymbol{C} = \left(\begin{array}{c|c} \boldsymbol{C}_{11} & \boldsymbol{O} \\ \boldsymbol{O} & \boldsymbol{C}_{22} \end{array}\right) \tag{6.5.19}$$

where $\boldsymbol{C}_{11}$ is an $(n - 1)$th-order non-singular matrix and $\boldsymbol{C}_{22}$ is a nth-order diagonal matrix $diag\left(\sqrt{M_1\omega_N/2}, \ldots, \sqrt{M_n\omega_N/2}\right)$.

Differentially (6.5.18), one obtains:

$$\begin{cases} \partial H/\partial \boldsymbol{Y} = 2\boldsymbol{Y} \\ \partial^2 H/\partial \boldsymbol{Y}^2 = 2\boldsymbol{E} \end{cases} \tag{6.5.20}$$

where $\boldsymbol{E}$ is a $(2n - 1)$th-order unit matrix.

Using $\boldsymbol{Y} = \boldsymbol{CX}$ and $\boldsymbol{Q} = \boldsymbol{C}^T\boldsymbol{C}$, (6.5.17) is transformed as follows:

$$\mathrm{d}\boldsymbol{Y} = \boldsymbol{C}\mathrm{d}\boldsymbol{X} = 2\boldsymbol{CMC}^T\boldsymbol{Y}\mathrm{d}t + 2\boldsymbol{CDC}^T\boldsymbol{Y}\mathrm{d}t + \boldsymbol{CG}\mathrm{d}\boldsymbol{B}(t). \tag{6.5.21}$$

Based on Itô formula, Itô equation of system energy (6.5.18) can be expressed as follows:

$$\begin{aligned} \mathrm{d}H &= (\partial H/\partial \boldsymbol{Y})^T\mathrm{d}\boldsymbol{Y} + (1/2)tr(\boldsymbol{CGG}^T\boldsymbol{C}^T\partial^2 H/\partial \boldsymbol{Y}^2)\mathrm{d}t \\ &= 4\boldsymbol{Y}^T\boldsymbol{CDC}^T\boldsymbol{Y}\mathrm{d}t + tr(\boldsymbol{CGG}^T\boldsymbol{C}^T)\mathrm{d}t + 2\boldsymbol{Y}^T\boldsymbol{CG}\mathrm{d}\boldsymbol{B}(t) \end{aligned} \tag{6.5.22}$$

where $tr(\boldsymbol{CGG}^T\boldsymbol{C}^T)$ is the trace of matrix $\boldsymbol{CGG}^T\boldsymbol{C}^T$ (i.e., the summation of all diagonal elements in $\boldsymbol{CGG}^T\boldsymbol{C}^T$).

One can derive the following equations easily:

$$\begin{cases} \boldsymbol{CDC}^T = diag[0,\ldots,0,-D_1/(2M_1),\ldots,-D_n/(2M_n)] \\ \boldsymbol{CGG}^T\boldsymbol{C}^T = diag[0,\ldots,0,\omega_N\sigma_1^2/(2M_1),\ldots,\omega_N\sigma_n^2/(2M_n)]\,. \\ \boldsymbol{CG} = diag[0,\ldots,0,\sigma_1\sqrt{\omega_N/(2M_1)},\ldots,\sigma_n\sqrt{\omega_N/(2M_n)}] \end{cases} \quad (6.5.23)$$

Combining (6.5.22) and (6.5.23), one obtains:

$$\begin{aligned} \mathrm{d}H &= \sum_{i=1}^{n}\left(\frac{-2D_i}{M_i}y_{n-1+i}^2 + \frac{\sigma_i^2\omega_N}{2M_i}\right)\mathrm{d}t + \sum_{i=1}^{n}\left[\sigma_i\sqrt{\frac{2\omega_N}{M_i}}y_{n-1+i}\mathrm{d}B_i(t)\right]. \\ &= \sum_{i=1}^{n}\left(\frac{-2D_i}{M_i}y_{n-1+i}^2 + \frac{\sigma_i^2\omega_N}{2M_i}\right)\mathrm{d}t + \sqrt{\sum_{i=1}^{n}\left(\frac{2\omega_N\sigma_i^2}{M_i}y_{n-1+i}^2\right)}\mathrm{d}B(t) \end{aligned} \quad (6.5.24)$$

According to the stochastic averaging, system energy H weakly converges to the first-order diffusion process as follows:

$$\mathrm{d}H = \bar{m}(H)\mathrm{d}t + \bar{\sigma}(H)\mathrm{d}B(t) \quad (6.5.25)$$

where

$$\begin{cases} \bar{m}(H) = \frac{1}{S}\int\limits_{\Omega}\sum_{i=1}^{n}\left(\frac{-2D_i}{M_i}y_{n-1+i}^2 + \frac{\sigma_i^2\omega_N}{2M_i}\right)\mathrm{d}s = \sum_{i=1}^{n}\left(\frac{\sigma_i^2\omega_N}{2M_i}\right) - \sum_{i=1}^{n}\left(\frac{2D_i}{M_i}\cdot\frac{\int_\Omega y_{n-1+i}^2\mathrm{d}s}{\int_\Omega 1\mathrm{d}s}\right) \\ \bar{\sigma}^2(H) = \frac{1}{S}\int\limits_{\Omega}\sum_{i=1}^{n}\left(\frac{2\omega_N\sigma_i^2}{M_i}y_{n-1+i}^2\right)\mathrm{d}s = \sum_{i=1}^{n}\left(\frac{2\omega_N\sigma_i^2}{M_i}\cdot\frac{\int_\Omega y_{n-1+i}^2\mathrm{d}s}{\int_\Omega 1\mathrm{d}s}\right) \\ S = \int\limits_{\Omega} 1\mathrm{d}s \\ \Omega = \left\{(y_1,\ldots,y_i,\ldots,y_{2n-1})|y_1^2+\ldots y_i^2\ldots+y_{2n-1}^2 = H\right\} \end{cases}.$$

$$(6.5.26)$$

Using the symmetry of the first type surface integral, one has

$$\int\limits_{\Omega} y_1^2\mathrm{d}s = \cdots = \int\limits_{\Omega} y_i^2\mathrm{d}s = \cdots = \int\limits_{\Omega} y_{2n-1}^2\mathrm{d}s. \quad (6.5.27)$$

According to the energy function (6.5.18), the following equation can be deduced as:

$$\sum_{i=1}^{2n-1}\int\limits_{\Omega} y_i^2\mathrm{d}s = \int\limits_{\Omega}\sum_{i=1}^{2n-1} y_i^2\mathrm{d}s = \int\limits_{\Omega} H\mathrm{d}s = H\int\limits_{\Omega} 1\mathrm{d}s. \quad (6.5.28)$$

Combining (6.5.28) and (6.5.27), one obtains:

$$\frac{\int_{\Omega} y_i^2 \mathrm{d}s}{\int_{\Omega} 1 \mathrm{d}s} = \frac{H}{(2n-1)}. \tag{6.5.29}$$

By incorporating (6.5.29) into (6.5.26), the following equations can be deduced as:

$$\begin{cases} \bar{m}(H) = \left(\sum_{i=1}^{n} \frac{\omega_N \sigma_i^2}{2M_i}\right) - \left[\sum_{i=1}^{n} \frac{2D_i}{M_i(2n-1)}\right] \cdot H \\ \bar{\sigma}^2(H) = \left[\sum_{i=1}^{n} \frac{2\omega_N \sigma_i^2}{M_i(2n-1)}\right] \cdot H \end{cases}. \tag{6.5.30}$$

In stochastic theory, the stability probability shown in (6.5.4) is defined as the reliability of diffusion process (6.5.10), which can be solved by Kolmogorov's backward equation as follows:

$$\frac{\partial P(t|H_0)}{\partial t} = \bar{m}(H_0)\frac{\partial P(t|H_0)}{\partial H_0} + \frac{\bar{\sigma}(H_0)^2}{2}\frac{\partial^2 P(t|H_0)}{\partial H_0^2}. \tag{6.5.31}$$

Equation (6.5.31) is a parabolic partial differential equation. A solution of a partial differential equation is not unique generally, so additional conditions must be specified on the boundary of the region where the solution is defined. The initial conditions and boundary conditions for (6.5.31) are as follows:

$$\begin{cases} \text{Initial condition}: \ \text{if } H_0 < H_{cr}, \ P(0|H_0) = 1 \\ \text{Boundary condition 1}: \ P(t|H_{cr}) = 0 \\ \text{Boundary condition 2}: \text{if } H_0 = 0, \ \frac{\partial P}{\partial t} = \bar{m}(0)\frac{\partial P}{\partial H_0} \end{cases} \tag{6.5.32}$$

where the initial condition denotes that stability probability P at $t = 0\ s$ equals 1, when the initial energy H_0 is lower than the critical energy H_{cr}; boundary condition 1 denotes that stability probability P is equal to 0, if the initial energy H_0 is higher than the critical energy H_{cr}; boundary condition 2 denotes that the solution at $H_0 = 0$ should satisfy $\frac{\partial P}{\partial t} = \bar{m}(0)\frac{\partial P}{\partial H_0}$, which is used to assure the solution with initial energy H_0 being higher than 0.

Several numerical calculation methods can be used to solve (6.5.31), such as the finite difference method. In this study, the finite difference method based on Crank Nicholson's scheme is applied.

To illustrate the solution of (6.5.31) clearly, Fig. 6.15 shows the discrete result mesh for Kolmogorov's backward equation. By the finite difference method, one considers $P(t|H_0)$ is only at discrete time values $t(i) = i\Delta t$ ($i = 0, 1, 2, \ldots$) and discrete initial energy values is $H_0(j) = j\Delta H_0$ ($j = 0, 1, 2, \ldots$). Then $t(i)$ and $H_0(j)$ constitute a discrete result mesh in the plane (t, H_0). Also, it can be seen that the initial conditions and boundary conditions define the region for the solution of (6.5.31).

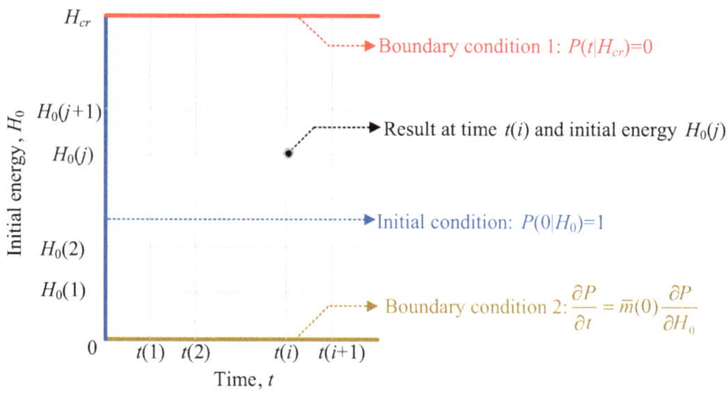

Fig. 6.15 The discrete result mesh of Kolmogorov's backward equation

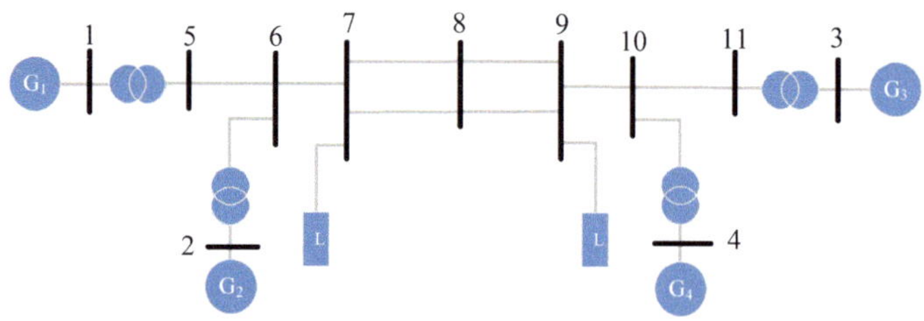

Fig. 6.16 Kundur's 4-machine 2-area power system

The Kundur's 4-machine 2-area system, which is shown in Fig. 6.16, is adopted as the first simulation system. The system is reduced to a 4-bus system by eliminating all the load buses. Then, the system is simulated with the integration of stochastic disturbance, which are represented as Gauss white noises with intensities of σ_1, σ_2, σ_3, and σ_4, at Generator 1, Generator 2, Generator 3 and Generator 4, respectively.

In this case study, the transient stability under stochastic disturbance is investigated by utilizing the proposed method, and results are compared with Monte Carlo simulation results. Three simulation cases with different disturbance intensity are demonstrated: Case 1: $\sigma_1 = \sigma_2 = \sigma_3 = \sigma_4 = 0.07$; Case 2: $\sigma_1 = \sigma_2 = \sigma_3 = {}_4 = 0.075$; Case 3: $\sigma_1 = \sigma_2 = \sigma_3 = \sigma_4 = 0.08$. Figure 6.17 shows the stability probability curves solved by the proposed method and Monte Carlo simulation. It can be seen that the analytical results are very close to Monte Carlo simulation results.

A 50-machine 145-bus test power system is also used to demonstrate the effectiveness of the proposed method. The system is reduced to a 50-bus system by eliminating load buses. Assume that all generators are subjected to stochastic disturbances with the same intensity. Three cases with different disturbance intensity are simulated: Case 1: small intensity; Case 2: medium intensity; Case 3: large

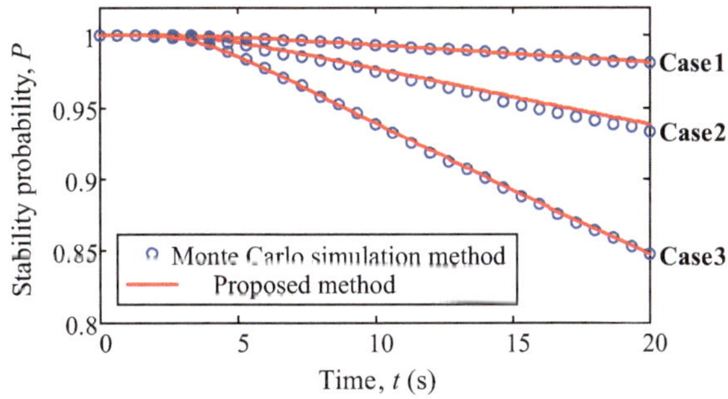

Fig. 6.18 Stability probability curves of a 50-machine 145-bus test power system under stochastic disturbance

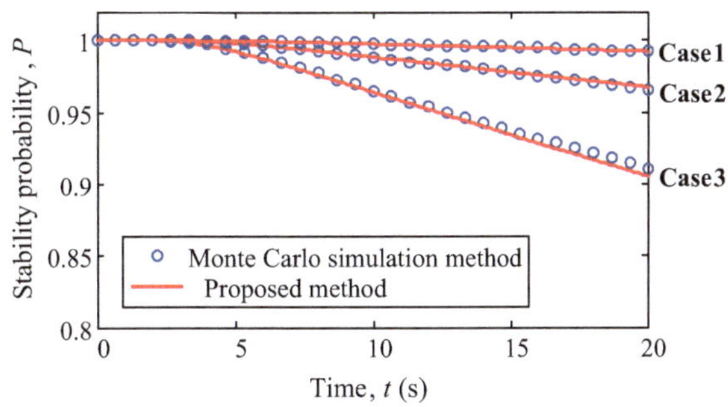

Fig. 6.17 Stability probability curves of Kundur's 4-machine 2-area power system under stochastic disturbance

intensity. Results from the proposed method and Monte Carlo simulation are shown in Fig. 6.18. It can be seen that analytical results are still close to Monte Carlo simulation results, which demonstrates the applicability of the proposed method in bulk power systems.

6.5.2 Stability Probability of Lossy System under Large Stochastic Disturbance

Recently, much attention has been attracted on the study of analyzing the stability in power systems under stochastic disturbance. However, the issue including the losses has not been addressed yet. To deal with the above issue, in this study, an analytical method is offered to assess the stability of power systems with losses under stochastic disturbance.

In this study, the stochastic model is initially presented for power systems with losses under stochastic disturbance, before the novel framework is proposed to analytically assess its stability. A major problem in the stability assessment by direct method is the derivation of Lyapunov functions for power systems that are with losses.

The losses are considered mainly from transmission systems and transfer conductances. In this study, a SMIB system with losses is underscored, whose model is as follows:

$$\begin{cases} \mathrm{d}\delta/\mathrm{d}t = \omega_{\mathrm{N}}\omega \\ M\mathrm{d}\omega/\mathrm{d}t = P - EUB_{12}\sin\delta - EUG_{12}\cos\delta - D\omega \end{cases} \tag{6.5.33}$$

where δ denotes the rotor angle of the machine with respect to infinite bus; ω denotes the rotating speed of the machine; M denotes the inertia coefficient of the machine; D denotes the damping coefficient of the machine; $P = P_m - G_{11}E^2$, and P_m denotes the mechanical power; E is the voltage behind direct axis transient reactance; G_{12} (G_{11}) denotes the conductance of the element in the first row and second (first) column of the reduced admittance matrix; B_{12} denotes the susceptance of the element in the first row and second column of the reduced admittance matrix; and ω_{N} denotes the synchronous machine speed.

The losses from the transmission systems and transfer conductances are expressed by $EUG_{12}\cos\delta$. Usually, the model (6.5.33) can be expressed in the more compact form:

$$\begin{cases} \mathrm{d}\delta/\mathrm{d}t = \omega_{\mathrm{N}}\omega \\ M\mathrm{d}\omega/\mathrm{d}t = P - EUY_{12}\sin(\delta - \alpha) - D\omega \end{cases} \tag{6.5.34}$$

where $Y_{12}^2 = B_{12}^2 + G_{12}^2$ and $\alpha = -\arctan(G_{12}/B_{12})$.

There exists the phase α in (6.5.34), because of the losses. Due to the presence of losses in the model (6.5.33) and (6.5.34), this kind of system is called as the lossy power system.

Practically, the losses are neglected in the lossy power system when the direct method is used to assess the stability, due to no existence of the general energy function in the lossy power systems generally. However, for the SMIB system (6.5.34), there exists the general energy function, which is shown as follows:

$$H = \frac{1}{2}M\omega_{\mathrm{N}}\omega^2 - P\delta - EUY_{12}\cos(\delta - \alpha) + P\delta_s + EUY_{12}\cos(\delta_s - \alpha) \tag{6.5.35}$$

where δ_s is the stable state of δ.

Based on the energy function in the direct method, the stability can be easily assessed. To be specific, if the system energy (6.5.35) exceeds the critical potential energy, the system is not stable. For the system (6.5.34) with the energy function (6.5.35), the critical potential energy is equal to the potential energy when $\delta = -\delta_s + 2\alpha$. The potential energy is shown as follows:

$$H_P = -P\delta - EUY_{12}\cos(\delta - \alpha) + P\delta_s + EUY_{12}\cos(\delta_s - \alpha). \tag{6.5.36}$$

In this study, the stochastic model is initially presented for the power systems with losses under stochastic disturbance, before the novel framework is proposed to analytically assess its stability. To describe the power systems with losses under stochastic disturbance, stochastic differential equations are adopted in this study to express the stochastic model. The normalized SMIB model with additive white noise perturbations is modeled as a set of Itô stochastic differential equations.

$$\begin{cases} \mathrm{d}\delta/\mathrm{d}t = \omega_{\mathrm{N}}\omega \\ M\mathrm{d}\omega/\mathrm{d}t = P - EUY_{12}\sin(\delta - \alpha) - D\omega + \sigma W(t) \end{cases} \tag{6.5.37}$$

where $\sigma W(t)$ denotes the stochastic disturbance, which represents the power unbalance between the generators and the loads; $W(t)$ is the Gaussian white noise; σ is the intensity of the excitation.

Because the network and load are merged together, the term $G_{11}E^2$ in P actually denotes the active power at load locations. In other words, the stochastic disturbance acting on machine equations describe the disturbances from not only the generation power but also the load power.

Based on the energy function method, the stability probability $P(t|H_0)$ can be expressed in the following form:

$$P(t|H_0) = P\{H(\tau) < H_{\mathrm{cr}}, \tau \in (0, t] | H(0) = H_0 < H_{\mathrm{cr}}\} \tag{6.5.38}$$

where $P(t|H_0)$ denotes the stability probability, which represents the probability that the system energy H (6.5.35) is smaller than the critical potential energy H_{cr} (6.5.35) within the time interval [0, t) when the initial system energy $H(0)$ is always smaller than the critical potential energy H_{cr}.

To express the power system model in Quasi-Hamiltonian form, the Hamiltonian should be defined first. For the model (6.5.33), the energy function of the power system can be regarded as the Hamiltonian. Furthermore, the stochastic model in Quasi-Hamiltonian form can be shown as follows:

$$\begin{cases} \mathrm{d}\delta/\mathrm{d}t = (1/M)\partial H/\partial\omega \\ \mathrm{d}\omega/\mathrm{d}t = (-1/M)\partial H/\partial\delta - [D/(M^2\omega_{\mathrm{N}})]\partial H/\partial\omega + (\sigma/M)W(t) \end{cases} \tag{6.5.39}$$

where

$$\begin{cases} \partial H/\partial\omega = M\omega_{\mathrm{N}}\omega \\ \partial H/\partial\delta = -P + EUY_{12}\sin(\delta - \alpha) \end{cases} \tag{6.5.40}$$

where ∂ denotes the partial differential operator.

If there are no stochastic disturbances in the model (6.5.37) (i.e., the deterministic model (6.5.34)), the differential function of the system energy could derivate as follows:

$$\mathrm{d}H = (\partial H/\partial\omega)\mathrm{d}\omega + (\partial H/\partial\delta)\mathrm{d}\delta \tag{6.5.41}$$

After the introduction of stochastic disturbance, the Itô stochastic differential equations of the system energy H can be obtained based on the Itô formula, which is shown as follows:

$$\begin{aligned} \mathrm{d}H &= (\partial H/\partial\omega)\mathrm{d}\omega + (\partial H/\partial\delta)\mathrm{d}\delta + [\sigma^2/(2M^2)](\partial^2 H/\partial\omega^2)\mathrm{d}t \\ &= [-D\omega_{\mathrm{N}}\omega^2 + \sigma^2\omega_{\mathrm{N}}/(2M)]\mathrm{d}t + \sigma\omega_{\mathrm{N}}\omega\mathrm{d}B(t) \end{aligned} \tag{6.5.42}$$

where $[\sigma^2/(2M^2)](\partial^2 H/\partial\omega^2)$ is the Wong-Zakai correction term; $B(t)$ is the Wiener process, and $\mathrm{d}B(t)/\mathrm{d}t = W(t)$.

Furthermore, by utilizing the SAM, the Hamiltonian (i.e., the system energy) weakly converges with a one-dimension diffusion process which can be regarded as the equivalent Itô stochastic differential equations for (6.5.42), as follows:

$$\mathrm{d}H = \bar{m}(H)\mathrm{d}t + \bar{\sigma}(H)\mathrm{d}B(t) \tag{6.5.43}$$

where

$$\begin{cases} \bar{m}(H) = \frac{1}{T(H)}\int\limits_{\Omega}\{[-D\omega_{\mathrm{N}}\omega^2 + \sigma^2\omega_{\mathrm{N}}/(2M)]/(\partial H/\partial\omega)\}\mathrm{d}\delta \\ \bar{\sigma}^2(H) = \frac{1}{T(H)}\int\limits_{\Omega}\left[(\sigma\omega_{\mathrm{N}}\omega)^2\Big/(\partial H/\partial\omega)\right] \times \mathrm{d}\delta \\ T(H) = \int\limits_{\Omega}[1/(\partial H/\partial\omega)] \times \mathrm{d}\delta \\ \Omega = \{(\delta)|H(\delta,0) \le H\} \end{cases} . \tag{6.5.44}$$

The final step in the proposed assessment method is the application of Kolmogorov's backward equation to solve for the stability probability.

After the stochastic averaging, the system energy of the complex stochastic system (6.5.39) is equivalent to the diffusion process (6.5.43). The stability probability (6.5.38) can be regarded as the first-passage probability of the diffusion process (6.5.43). In probability theory, Kolmogorov's backward equation is often used to characterize stochastic processes. To be specific, Kolmogorov's backward equation describes how the probability that a stochastic process is in a certain region changes over time. Based on the Kolmogorov's backward equation, the stability probability is calculated as the following equation:

$$\partial P(t|H_0)/\partial t = \bar{m}(H_0)\partial P(t|H_0)/\partial H_0 + \frac{1}{2}\bar{\sigma}(H_0)^2\partial^2 P(t|H_0)/\partial H_0^2 \tag{6.5.45}$$

where H_0 denotes the initial system energy (i.e., the system energy at $t = 0$).

For the partial differential Eq. (6.5.45), initial and boundary conditions should be given for the region where solutions of stability probability $P(t|H_0)$ will need to be solved.

Firstly, the initial system energy H_0 should be smaller than the critical system energy H_{cr}, the stability probability $P(t|H_0)$ at $t = 0$ is equal to 1, which represents the initial condition. If the initial system energy H_0 exceeds the critical energy H_{cr}, the stability probability $P(t|H_0)$ should be 0 all the time, which represents a boundary condition. Meanwhile, the stability probability $P(t|H_0)$ with initial energy H_0 should be higher than 0 all the time, which represents another boundary condition. These associated conditions for (6.5.45) are shown as follows:

$$\begin{cases} P(0|H_0) = 1, \text{ when } H_0 < H_{cr} \\ P(t|H_{cr}) = 0 \\ \partial P/\partial t = \bar{m}(0)\ \partial P/\partial H_0, \text{ when } H_0 = 0 \end{cases} \quad . \tag{6.5.46}$$

Usually, the partial differential equation based on Kolmogorov's backward Eq. (6.5.45) is hard to be solved directly. In this study, the Crank–Nicolson method is adopted to solve the function (10) to obtain stability probability.

In this section, a modified SMIB system is adopted to verify the proposed method. This system includes the losses and stochastic disturbance. The model of this system is described by (6.5.34), and the corresponding parameters are given as follow: $\omega_N = 314.16$, $M = 20$, $P = 1.1184$, $E = U = 1$, $Y_{12} = 2$, $\alpha = 6°$, $D = 05$, $\sigma = 0.1$, and $\delta_s = 40°$.

Based on the potential energy (6.5.36), the relation between the potential energy H_P and the rotor angle δ is shown in Fig. 6.19. It can be seen that the point A ($\delta = \delta_s = 34°$) is the stable equilibrium point, which represents the stable state of the system; the point B ($\delta = \pi - \delta_s - 2\alpha = 152°$) is the unstable equilibrium point, whose energy is the critical potential energy. Furthermore, the critical potential energy is calculated as $H_{cr} = H_P(152°) = 1.13$.

If the system energy exceeds the critical potential energy, the system is unstable. When the stochastic disturbance is introduced into the power system, it is a random event that system energy being less than the critical potential energy, which should be described by the probability. In this study, this kind of probability is defined as the stability probability, and the SAM is further used to analytically calculate it. The stability probability of the power system is equal to the first-passage probability (6.5.45) of the diffusion process (6.5.41).

After the SAM is utilized, the equivalent diffusion process can be obtained. Based on the analysis in this section, the coefficients (i.e., $\bar{m}(H)$ and $\bar{\sigma}(H)$) of the equivalent Itô stochastic differential equation for the system energy can be calculated, which are illustrated in Fig. 6.20.

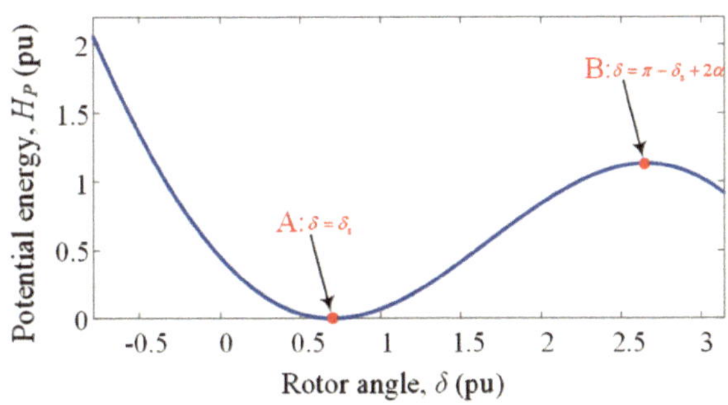

Fig. 6.19 Potential energy curve near the stable point

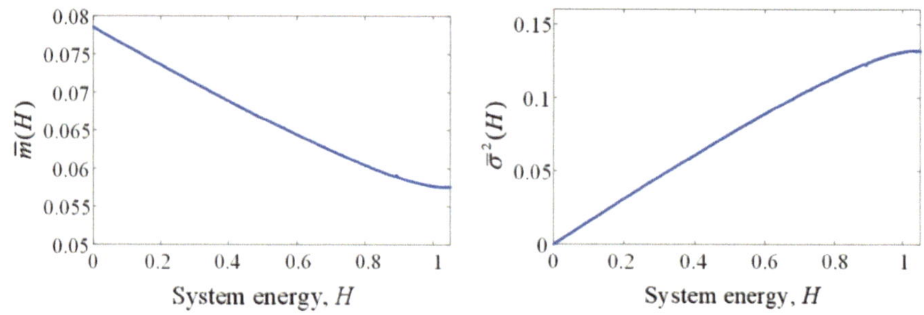

(a) The drift coefficient $\overline{m}(H)$ **(b) The diffusion coefficient** $\overline{\sigma}(H)$

Fig. 6.20 The coefficients of the equivalent Itô stochastic differential equation for the system energy

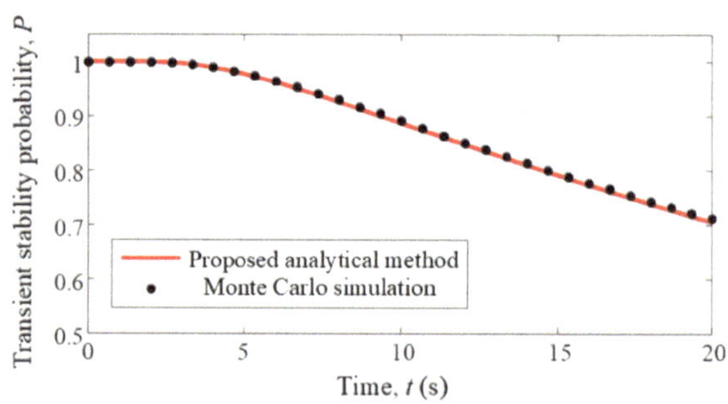

Fig. 6.21 Stability probability of power system with losses under stochastic disturbance

Furthermore, the stability probability is calculated based on (6.5.45), which is shown as the red line in Fig. 6.21. In this study, to verify the proposed method based on (6.5.45), Monte Carlo simulation is also utilized to calculate the stability probability, which is shown as the black dotted line in Fig. 6.21. As can be seen from Fig. 6.21, the analytical results agree well with those from the Monte Carlo simulation.

This analytical method also has a large advantage in time consumption. Based on the energy function method, the trajectories of system state should be calculated firstly, and then the system energy can be calculated based on the trajectories of system energy. Even though the stability is assessed by judging whether the system state trajectories are unstable directly, iteration which costs most time is still necessary for Monte Carlo simulation. In Monte Carlo simulation, system states trajectories are simulated by repeated trials based on the stochastic model of power system with losses under stochastic disturbance (6.5.33), which is the major reason that much computation is needed in Monte Carlo simulation.

From these respects, the proposed analytical method is better, due to its higher efficiency and clearer mechanism. Moreover, the proposed method is analytical, which can offer the accurate results directly without iterations. Hence, the proposed method could work well for analyzing the stochastic stability in power systems with losses under stochastic disturbance.

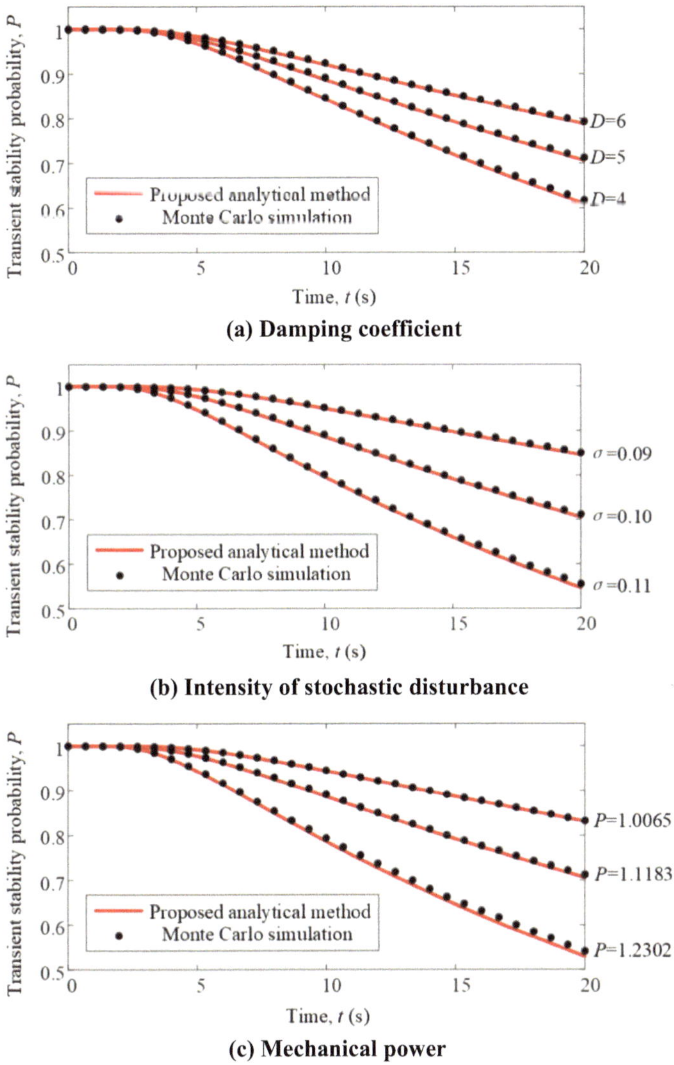

Fig. 6.22 Stability probability of power system with losses under stochastic disturbance

To study the impacts of parameters on the stability probability in the power systems with losses under stochastic disturbance, several simulation cases with different damping coefficient, the intensity of stochastic disturbance, and mechanical power (which is included in P), whose results are shown in Fig. 6.22. When the mechanical power varies, the system energy function and the critical energy change accordingly. From Fig. 6.22, it can be seen that the results from the proposed method agree well with those from Monte Carlo simulation. Moreover, the stability probability decreases with the decreasing of the damping coefficient, increasing of the intensity of stochastic disturbance, and increasing of the mechanical power. These results suggest that the operations on enlarging system damping, mitigating stochastic disturbance and reducing the generation level could improve the stability probability.

6.5.3 Stability Probability of Power System under both Stochastic Disturbance and Fault

In the transient stability analysis, researchers tend to concern about whether the system can stay in the attraction domain of the stable equilibrium point under the fault, that is, whether the system energy H is less than the critical energy H_{cr}. Assume that the system runs at the stable equilibrium point before the fault, and the system energy is 0. During the fault, the power balance is broken, so the system leaves the equilibrium point and the system energy jumps sharply from 0 to H_0. After the fault is removed, the system will moves under the stochastic disturbance with the initial energy H_0, which is a stochastic process. The changes to the system energy H are as shown in Fig. 6.23, when three sampling curves are given after the fault is removed.

In the transient process, the kinetic energy and the potential energy interchange. The stability of the system depends on the ability of the system to absorb transient energy after the fault, that is, whether the system can reduce the kinetic energy to zero before reaching the dynamical saddle point, so as to stay in a stable domain, and the maximum energy that the system can absorb is the critical energy H_{cr}.

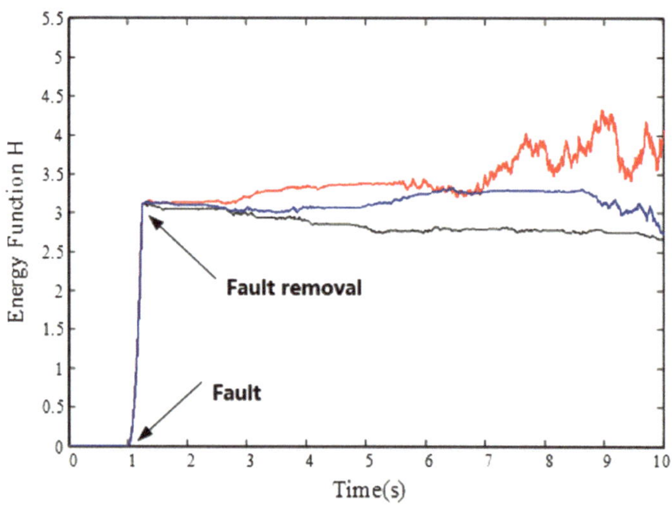

Fig. 6.23 Schematic diagram of the changes to the energy function

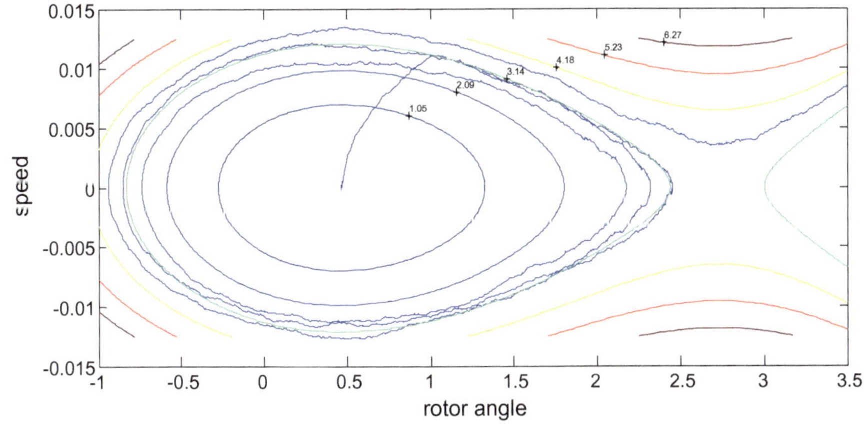

Fig. 6.24 Phase trajectory of stochastic power system after the fault

Although the H_{cr} for an actual system can only be obtained through approximate calculation, it is still an important indicator to measure the transient stability of the system.

For a stochastic power system, a phase trajectory of the system states after the fault is shown in Fig. 6.24. Due to the effects of a damping force, a restoring force, and a stochastic disturbance, the transient energy fluctuates continuously. That is, the phase trajectory drifts among different equal-energy curves. And if the system's transient energy is greater than H_{cr} at a certain point, the system is considered to be unstable. The phase trajectory shown in Fig. 6.24 is the result of only one sampling where the system eventually loses its stability in this simulation. This result is stochastic, so it is possible to draw very different conclusions under different samplings.

For a stochastic power system, the conditional reliability can be used to describe its probabilistic stability. For a stochastic system, the conditional reliability of the system can be defined as

$$R(t|H_0) = \mathrm{P}\{H(s) \in [H_{\min}, H_{cr}),\ \ \mathrm{s} \in (0, t]|H_0 \in [H_{\min}, H_{cr})\} \tag{6.5.47}$$

$R(t|H_0)$ describes the probability that the system is still in the stability region at time t after the fault when the initial energy of the system is H_0. According to the stochastic dynamics theory, $R(t|H_0)$ satisfies the Kolmogorov's backward equation:

$$\frac{\partial R}{\partial t} = \bar{m}(H_0)\frac{\partial R}{\partial H_0} + \frac{1}{2}\bar{\sigma}^2(H_0)\frac{\partial^2 R}{\partial H^2} \tag{6.5.48}$$

This equation is a parabolic partial differential equation where an analytical solution cannot be obtained in general. It needs to be simplified.

For a stochastic power system, the EEAC method can be used to obtain a simplified system described by Eq. (6.3.9). If the coefficient l_i of the stochastic disturbance of the system is independent of the state variables, the Wong-Zakai correction term of the system (6.3.9) is zero, and it is converted into an Itô stochastic differential equation according to the method described in Sect. 6.3.2:

$$\begin{bmatrix} \mathrm{d}\delta_{SA} \\ \mathrm{d}\omega_{SA} \end{bmatrix} = \left\{ J \begin{bmatrix} \frac{\partial H}{\partial \delta_{SA}} \\ \frac{\partial H}{\partial \omega_{SA}} \end{bmatrix} + d \begin{bmatrix} \frac{\partial H}{\partial \delta_{SA}} \\ \frac{\partial H}{\partial \omega_{SA}} \end{bmatrix} \right\} \mathrm{d}t + \sigma \mathrm{d}B_s(t) \tag{6.5.49}$$

where $J = \begin{bmatrix} & -\frac{1}{M} \\ \frac{1}{M} & \end{bmatrix}$ is a structure matrix; $d = \begin{bmatrix} 0 & 0 \\ 0 & -\frac{k}{M\omega_N} \end{bmatrix}$ is a damping matrix; $B_s(t)$ is a standard Wiener process; $\sigma\sigma^T = 2l\zeta l^T$. One can derive from Eq. (6.5.3) that:

$$\mathrm{d}H = \bar{m}(H)\mathrm{d}t + \bar{\sigma}(H)\mathrm{d}B_s(t) \tag{6.5.50}$$

where $\bar{m}H = \frac{1}{T}\int_\Omega \left(-k_{SA}M\omega_{SA} + \frac{1}{2}\sigma^2 M \frac{1}{\omega_{SA}}\right)\mathrm{d}\sigma_{SA}$; $\bar{b}H = \bar{\sigma}^2(H) = \frac{1}{T}\int_\Omega \sigma^2 M^2 \omega_N \omega_{SA} \mathrm{d}\sigma_{SA}$; $T = \int_\Omega \frac{1}{\omega_N \omega_{SA}}\mathrm{d}\sigma_{SA}$; $\Omega = \{\omega_{SA} | H(\delta_{SA}, \omega_{SA}) \le H\}$. The Eq. (6.5.49) describes the diffusion law of the equivalent system Hamilton energy. According to the diffusion equation theory, the corresponding Kolmogorov's backward equation can be directly solved to obtain the conditional security and reliability of the stochastic power system.

The quasi Hamilton system SAM (QHSSAM) is applied to the four-machine two-area disturbed system as shown in Fig. 6.25. It is assumed that the generator 4's input mechanical power is stochastic, and the disturbance follows Gauss normal distribution $N(0, \sigma^2)$. $\sigma_4 = 1$, the disturbance coefficient is $f_{44} = 0.08$, and the generator damping coefficient is $D_i = 0.01$. The parameters of the generator, line, and load are detailed in [7]. Assume $t_f = 0.1$ s and three-phase short-circuit to ground faults occur on the bus 6. The transient stability of the system under different fault removal times t_{cl} is analyzed using the QHSSAM.

Such simulation shows that generators G1 and G2 in the four-machine two-zone system have good coherence, which can be regarded as S group, while G3 and G4

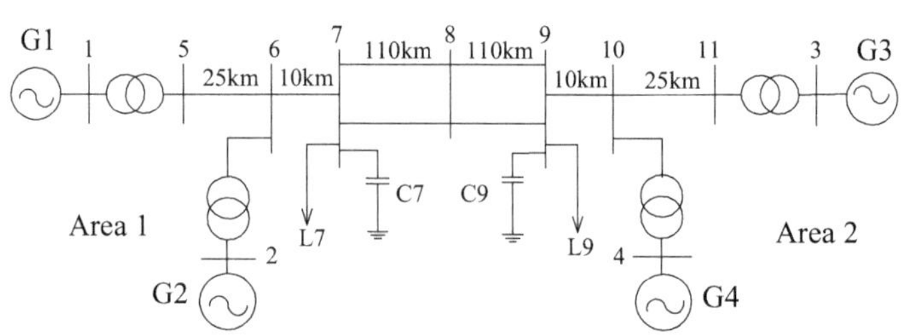

Fig. 6.25 4-machine 2-area system schematic

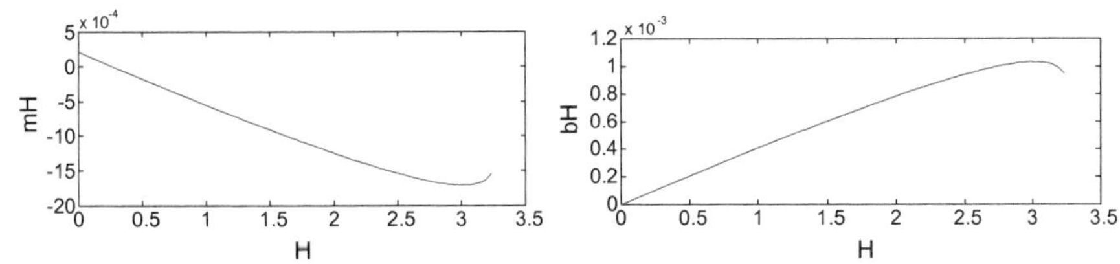

(a) Average drift coefficient **(b) Average diffusion coefficient**

Fig. 6.26 Average drift coefficient and diffusion coefficient for the diffusion equation

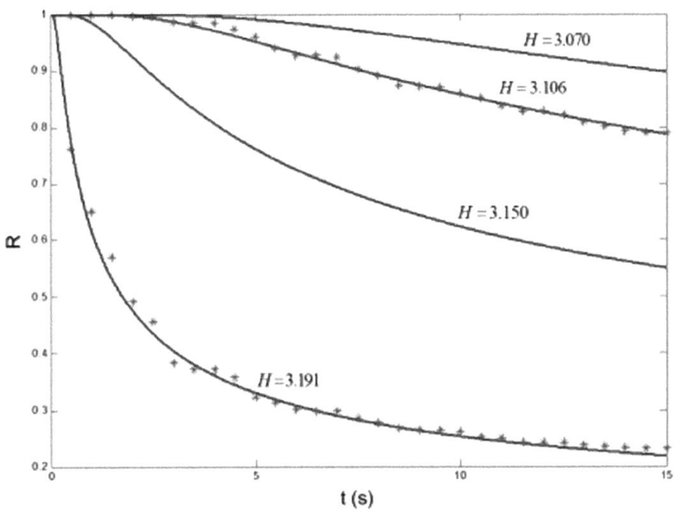

"---" for QHSSA method analysis results,
"*" for Monte Carlo simulations (3000 samplings)

Fig. 6.27 System conditional reliability curve

can be regarded as group A, and the system safe operating domain is $H \in [0, 3.219)$. Diffusion equation coefficients are calculated by Eq. (6.5.48), and the average drift coefficients $\bar{m}H$ and diffusion coefficients $\bar{b}H$ are as shown in Fig. 6.26.

The corresponding Kolmogorov's backward equation is solved by using the finite difference method, and the system conditional reliability is shown in Fig. 6.27. To verify the analysis accuracy of the QHSSAM, the Monte Carlo simulations are carried out upon the example system under the conditions of H_0=3.106 ($t_{cl} = 0.35\,\mathrm{s}$) and H_0=3.191 ($t_{cl} = 0.353\,\mathrm{s}$). By sampling for 3000 times, the statistical results are shown in Fig. 6.27. As seen from the figure, the analytical solutions are highly consistent with the Monte Carlo simulation results.

As shown in Fig. 6.27, as the initial energy H_0 increases, the security and reliability of the system decrease after the fault. Therefore, the security and reliability requirements of the disturbed system can be pre-set as $R(t|H_0) > \varepsilon$. That is, the probability that the system stays in the security domain is greater than ε at the point t after the system fault. Furthemore, the maximum H_0 that the system can withstand can be determined by referring to Fig. 6.27, and the required time for the removal of the fault can be calculated.

Conclusion: (1) The analytical solutions of the intra-region probability to power systems under small stochastic disturbances are derived. (2) The analytical solutions of the intra-region probability to power systems under large stochastic disturbances are derived. (3) The estimated values for the oscillation amplitudes of power systems under small stochastic disturbances are given. The above results can be used to obtain the security indexes of power systems under stochastic disturbances. However, the following question is, if these indexes cannot meet the requirements, how to take control measures to meet the requirements. This question is the content of the next chapter.

References

1. Stratonovich, R.L. 1963. *Topics in the Theory of Random Noise*. New York: Gordon Breach.
2. Khasminskii, R.Z. 1966. A limit theorem for the solution of differential equations with random right-hand sides. *Theory of Probability and Its Application* 11: 390–406.
3. Zhu, Weiqiu. 2003. *Nonlinear Stochastic Dynamics and Control—Hamiltonian Theoretical Framework*. Beijing: Science Press. (in Chinese).
4. Zhu, Weiqiu, Z.L. Huang, and Y. Suzuke. 2001. Response and stability of strongly nonlinear oscillators under wide-band random disturbance. *International Journal of Non-Linear Mechanics* 36: 1235–1250.
5. Zhu, Weiqiu and Guoqiang Cai. 2017. *Introduction to Stochastic Dynamics*. Beijing: Science Press. (in Chinese).
6. Xue, Yusheng. 1999. *Quantitative Theory for Motion Stability*. Nanjing: Jiangsu Science and Technology Press. (in Chinese).
7. Kurdur, P. 1994. *Power System Stability and Control*. New York: McGraw-Hill.

Chapter 7
Stochastic Optimal Control of Power System

Abstract It is hoped that the dynamic security of the power system under stochastic disturbances is as high as possible, but it comes at a cost. To get the best dynamic security at the lowest cost, stochastic optimal control is the best choice. In this chapter, stochastic control methods based on power control and excitation control are proposed for stochastic dynamic security.

7.1 Introduction

The purpose of studying stochastic power system is how to control it so that the power system runs according to the expected target. The so-called optimal control problem is to choose control input variables manually so that an optimized performance index of the power system reaches maximum or minimum. The stochastic optimal control theory mainly studies the Markov feedback control of the diffusion process. The control object is modeled as diffusion process, which is described by Itô stochastic differential equation. Based on the latest available information about the state of the system, the controller selects the optimal control from all possible controls that satisfy the constraint conditions, so that the system under the control reaches the optimal desired result of the intended target. At present, stochastic optimal control theory is mainly used in economics, especially financial issues, and it has a wide application prospect in the fields of physics, biology, engineering, management and other fields of science [1–10].

The two main methods to solve the problems of deterministic and stochastic optimal control are Pontryagin maximum principle and Bellman dynamic programming, which are necessary conditions for optimal control. Under certain conditions, they are also sufficient conditions [4].

The maximum principle was proposed by Pontryagin and his team in the 1950s, as a milestone in optimal control. According to the principle, any optimal control along with the state trajectory of the optimal system can be obtained by solving the augmented Hamiltonian equation set, which includes the original controlled system equations and their initial conditions, the adjoint equations and their final

P. Ju, *Stochastic Dynamics of Power Systems*, Power Systems,
https://doi.org/10.1007/978-981-13-1816-0_7

conditions, and the Hamiltonian maximum conditions. It is a forward-backward deterministic or stochastic ordinary differential equation set, and is about an issue of initial value-final value. The mathematical meaning of maximum principle lies in that it turns the optimal control problem of an infinite dimension into a maximum problem of a much simpler function.

The basic idea of dynamic programming proposed by Bellman in the 1950s is to consider a set of optimal control problems with different initial moments and initial conditions, contact them by applying the dynamic programming equation (Hamilton-Jacobi-Bellman equation, HJB equation for short). In this way, an optimal feedback control law is determined from the corresponding Hamiltonian function therein by taking the minimum or maximum conditions. The optimal control law is substituted into the dynamic programming equation to obtain the final dynamic programming equation, and then solve the final dynamic programming equation for the optimal control law in turn. The classical dynamic programming method requires that the dynamic programming equation has a classical solution, that is, a sufficiently smooth solution. However, for the problem of deterministic optimal control, there are no classical solutions to dynamic programming equations when the diffusion matrix is singular. In the early 1980s, Crandall and Lions introduced the so-called viscous solution for the first time, making dynamic programming method as a powerful tool for solving optimal control problems. The viscous solution mentioned here refers to the continuous but non-smooth solution of the partial differential equation. Its key feature is the use of super/sub-differential to replace the ordinary differential so that the uniqueness of solution can be maintained under proper conditions.

Historically, the maximum principle and dynamic programming have been developed independently. However, they are two different approaches to solving the same problem, like the Hamiltonian canonical equation and Hamilton-Jacobi (HJ) equation in classical mechanics. The extended Hamiltonian equation set in the maximum principle and the HJB equation in dynamical programming are equivalent to each other. The solution of the former can be directly expressed by the so-called four-step method using the latter. On the contrary, the solution of the latter can be expressed by the generalized Feynman-Kac formula using the former.

For the stochastic optimal control problem, the dynamic programming method is more effective because there have been many research results on the theory and numerical solution to the HJB equation [4]. However, the study on the solution to forward-backward stochastic differential equation set is still in its infancy.

7.2 Stochastic Optimal Control Principle

The specific formulation of the stochastic optimal problem depends on the type of the motion equation of the controlled system, the constraints imposed on the control, the performance indexes, the control time interval, and so on [4].

7.2.1 Motion Equation of Controlled System

The motion equation of the controlled system is the Itô stochastic differential equation as follows [4]:

$$\begin{cases} d\mathbf{X}(t) = \mathbf{m}(\mathbf{X}, \mathbf{u}, t)dt + \boldsymbol{\sigma}(\mathbf{X}, \mathbf{u}, t)d\mathbf{B}(t), t \in [t_0, t_f] \\ \mathbf{X}(t_0) = \mathbf{X}_0 \end{cases} \tag{7.2.1}$$

In the equation, $\boldsymbol{X}(t)$ is the state process of n-dimensional vector system; $\boldsymbol{B}(t)$ is the standard wiener process of m-dimensional vector; $\boldsymbol{u}(t)$ is the function $\boldsymbol{u}(\boldsymbol{X}(t)t)$ of the state $\boldsymbol{X}(t)$ of the same point, which is the r-dimensional vector Markov feedback control process; t_0 is the initial time, and it often takes $t_0 = 0$ when $\boldsymbol{m}$ and $\boldsymbol{\sigma}$ do not contain t; t_f is the final control time, which can be finite value, infinite value or stochastic variable; $\boldsymbol{m}$ and $\boldsymbol{\sigma}$ are given n-dimensional vector function and $n \times m$-dimensional matrix function, respectively, to satisfy the existence and uniqueness conditions.

Equation (7.2.1) is the general form of controlled system equation, which is often involved in the following special cases when applied:

1. $\boldsymbol{\sigma}$ does not contain control $\boldsymbol{u}$;
2. $\boldsymbol{m}$ and $\boldsymbol{\sigma}$ do not contain time t;
3. $\boldsymbol{m}$ and $\boldsymbol{\sigma}$ are the liner function of $\boldsymbol{X}(t)$;
4. Some combination of the above cases.

7.2.2 Control Constraint

Control is often subject to some constraint, the form of which depends on the controller. One of its constraint forms is:

$$\mathbf{u} \in U,\ U \subset R^r \tag{7.2.2}$$

It shows that all the samples of the control process belong to the set U, which can be interpreted as the limitation to the control force. Another possible constraint form is:

$$E\left[\int_{t_0}^{t_f} |\mathbf{u}(t)|^{\gamma} dt\right] \le u_0 < \infty \tag{7.2.3}$$

In the equation, $\gamma > 0$, $u_0 > 0$ is a constant, $E[\cdot]$ is the expected operator. When $\gamma = 1$, Eq. (7.2.3) can be interpreted as a limit on the average total control momentum. When $\gamma = 2$, it can be interpreted as a limit on the average total control energy. There are other forms of control constraints as well. Control that satisfies the conditions of control constraint can be called the feasible control. The feasible control which makes (7.2.1) have a unique solution is called the admissible control.

7.2.3 Performance Index

The objective of optimal control is usually expressed as the minimum or maximum of a functional, which is called the cost functional or performance index. For the stochastic optimal control, the state and control of the controlled system are all stochastic processes, and the functional is a stochastic variable, so the performance index is taken as the mathematical expectation of the functional [4].

Regarding the control issue over a fixed finite time interval, the performance index usually takes the form of

$$J(\mathbf{u}) = E\left[\int_{t_0}^{t_f} f(\mathbf{X}, \mathbf{u}, t)\mathrm{d}t + g\left(\mathbf{X}\left(t_f\right)\right)\right] \tag{7.2.4}$$

In the equation, f and g are given functions, which are respectively called runnning cost and the terminal cost. When f and g are both non-zero, the control problems (7.2.1) and (7.2.4) are called Boltz problems; they are Lagrange problems when it comes to $f \neq 0$ and $g = 0$, or Mayer problems when it comes to $f = 0$ and $g \neq 0$. Letting

$$\begin{aligned} \dot{X}_{n+1} &= f(\mathbf{X}, \mathbf{u}, t) \\ X_{n+1}(t_0) &= 0 \end{aligned} \tag{7.2.5}$$

should turn Boltz problems into Mayer problems. At this point, the motion equation of the controlled system consists of (7.2.1) and (7.2.5) with a dimension of $n + 1$, and the new performance index is:

$$J_1(\mathbf{u}) = E\left[g\left(\mathbf{X}\left(t_f\right)\right) + X_{n+1}\left(t_f\right)\right] \tag{7.2.6}$$

For the stochastic optimal control problem in a semi-infinite long time interval, the discount cost function can be considered, and only ergodic control is considered here. For a system that can reach smooth ergodic traversal state after a long time, for example, $\boldsymbol{m}, \boldsymbol{\sigma}$ or $\boldsymbol{u}$ do not contain t or the periodic or almost periodic function of t, the final steady-state distribution has nothing to do with the initial state of the controlled system, and the following long-run average performance index can be considered.

$$J(\mathbf{u}) = \lim_{t_f \to \infty} \frac{1}{t_f} \int_0^{t_f} f(\mathbf{x}, \mathbf{u}, t)\mathrm{d}t \tag{7.2.7}$$

The corresponding control problem is called the ergodic control.

For a controlled system defined on a finite field $D \subset R^n$, the upper limit of the integration as to performance index can be the first time the system has crossed the D

$$\tau = \inf(t \geq 0, \mathbf{X}(t) \notin D) \tag{7.2.8}$$

inf means the lower bound, and it can be replaced by min when the value can actually be reached. The corresponding performance index form is:

$$J(\mathbf{u}) = E\left[\int_{t_0}^{\tau} f(\mathbf{X}, \mathbf{u}, t)\mathrm{d}t + g(\mathbf{X}(\tau))\right] \tag{7.2.9}$$

When $f = 1$ and $g = 0$, Eq. (7.2.9) means the average time for the first time it makes cross

$$J(\mathbf{u}) = E[\tau(\mathbf{X}, \mathbf{u})] \tag{7.2.10}$$

The performance index can also be the probability that the system stays in the finite field D, i.e. the reliability function

$$J(\mathbf{u}) = P(\mathbf{X}(t) \in D, t_0 \leq t \leq \tau) \tag{7.2.11}$$

Although there are other forms of performance indexes, the main considerations in this book are (7.2.4), (7.2.7), (7.2.10) and (7.2.11).

The problem of stochastic optimal control is to choose a control in the admissible control set and to achieve the best performance index under the condition of (7.2.1), for example:

$$J(\mathbf{u}^*) = \inf_{\mathbf{u} \in U} J(\mathbf{u}) \tag{7.2.12}$$

In the equation, $\mathbf{u}^*$ is called the optimal control. The system state generated by the optimal control, that is, the solution $\mathbf{X}^*(t)$ of (7.2.1), is called the optimal state or trajectory, and $\mathbf{u}^*$ and $\mathbf{X}^*(t)$ constitute the optimal pair.

As Itô stochastic differential equations have both strong solutions and weak solutions, the stochastic optimal control should have both strong and weak formulations as well. The solution to the problem of dynamic programming is essentially weak formulation of the stochastic optimal control because the objective of stochastic optimal control is to minimize the mathematical expectation of a stochastic variable depending only on the probability distribution of the involved process instead of its trajectory. For most stochastic control problems, the weak formulation is enough.

7.2.4 Stochastic Dynamic Programming Method

The stochastic dynamic programming method is to establish and solve the stochastic dynamic programming equation for a given stochastic optimal control problem, determine the optimal control, and then solve the optimal state [4].

(1) **Value Function**

The value function, also known as the Bellman function, or the optimal cost function, refers to the minimum or the maximum of the performance index (or cost functional) as a function of the initial time and the initial state, which is a tool for analyzing optimal control problems using the dynamic programming method. It has different meanings for different control issues [4].

Consider a set of stochastic optimal control problems with different initial moments and different initial states, including the motion equation of the controlled system:

$$\begin{aligned} &\mathrm{d}\mathbf{X}(t) = m(\mathbf{X},\mathbf{u},s)\mathrm{d}s + \boldsymbol{\sigma}(\mathbf{X},\mathbf{u},s)\mathrm{d}\mathbf{B}(s) \\ &\mathbf{X}(t) = \mathbf{x} \end{aligned} \tag{7.2.13}$$

and performance index:

$$J(\mathbf{u}) = E\left[\int_t^{t_f} f(\mathbf{X},\mathbf{u},t)\mathrm{d}t + g\left(\mathbf{X}\left(t_f\right)\right)\right] \tag{7.2.14}$$

The solution of (7.2.13) resulting from the feedback control $\boldsymbol{u} = \boldsymbol{u}(\boldsymbol{X},s)$ in the initial state $\boldsymbol{X}_{\mathrm{tx}}(\boldsymbol{u},t) = \boldsymbol{x}$ is written as $\boldsymbol{X}_{\mathrm{tx}}(\boldsymbol{x},s)$. Under the control constraint (7.2.2), the value function is defined as:

$$V(\mathbf{x},t) = \inf_{\mathbf{u}\in U} E\left[\int_t^{t_f} f(\mathbf{X}_{tx},\mathbf{u},s)\mathrm{d}s + g\left(\mathbf{X}_{tx}\left(t_f\right)\right)\right] \tag{7.2.15}$$

In the equation, $\inf\limits_{\mathbf{u}\in U}$ means to take a minimum value for all the controls in U, and the expectation operator E should be understood as the one (E_{tx}) under the following conditions:

$$E_{tx}[F(\mathbf{Y})] = \int_D F(\mathbf{y})p(\mathbf{y},s|\mathbf{x},t)\mathrm{d}\mathbf{y} \tag{7.2.16}$$

Wherein $p(\boldsymbol{y},s|\boldsymbol{x},t)$ is the transition probability density of diffusion process $\boldsymbol{X}_{\mathrm{tx}}(\boldsymbol{u}, s)$. At the end of control, the value of the value function is:

$$V(\mathbf{x}, t_f) = g(\mathbf{X}_{tx}(t_f)) \tag{7.2.17}$$

$\boldsymbol{X}_{\mathrm{tx}}$ in (7.2.15) and (7.2.17) is often abbreviated as $\boldsymbol{X}$.

For other stochastic optimal control problems, the value function can be defined similarly. For example, for the optimal control problem of the maximum average first passage time (7.2.10), for the controlled system (7.2.1) under the control constraint (7.2.2), the value function is defined as:

$$V(\mathbf{x}, t) = \sup_{\mathbf{u} \in U} E[\tau_{tx}(\mathbf{X}_{tx}, \mathbf{u})] \tag{7.2.18}$$

(2) Establishment of Stochastic Dynamic Programming Equation

The stochastic dynamic programming equation represents the necessary conditions for stochastic optimal control. It is derived from the principle of dynamic programming (Bellman's Optimality Principle). The formal derivation of the dynamic programming equation of stochastic optimal control problem for (7.2.1), (7.2.2) and (7.2.4) is given below first, and then the dynamic programming equations of other stochastic optimal control problems are briefly described.

According to the dynamic programming principle, if $\boldsymbol{u}^*$ is the optimal feedback control of an optimal control problem over the entire time interval $[t, t_{\mathrm{f}}]$, $\boldsymbol{u}^*$ has the following properties: no matter what admissible control $\boldsymbol{u}$ is to be taken at the moment $t + h \in [t, t_{\mathrm{f}}]$ or over the time interval $[t, t_{\mathrm{f}}]$, $\boldsymbol{u}^*$ must necessarily be the optimal control of the state $\boldsymbol{X}_{\mathrm{tx}}(\boldsymbol{u}, t + h)$ produced from control $\boldsymbol{u}$ over $[t, t + h]$ wherein the same optimal control problem lies in the time interval $[t + h, t_{\mathrm{f}}]$.

For the stochastic optimal control problem (7.2.1), (7.2.2) and (7.2.4), and the value function definition (7.2.15), the value function over the entire interval $[t, t_{\mathrm{f}}]$ is recorded as $V(\boldsymbol{x},t)$, and take the value function over the interval $[t + h, t_{\mathrm{f}}]$ as $V(x(t + h), t + h)$. According to the above dynamic programming principle, there is:

$$\begin{aligned}
V(x,t) &= \inf_{\mathbf{u} \in U} E\left[\int_t^{t_f} f(\mathbf{X}, \mathbf{u}, s)\mathrm{d}s + g(\mathbf{X}(t_f))\right] \\
&= \inf_{\mathbf{u} \in U} E\left[\int_t^{t+h} f(\mathbf{X}, \mathbf{u}, s)\mathrm{d}s + \int_{t+h}^{t_f} f(\mathbf{X}, \mathbf{u}, s)\mathrm{d}s + g(\mathbf{X}(t_f))\right] \\
&= \inf_{\mathbf{u} \in U} E\left[\int_t^{t+h} f(\mathbf{X}, \mathbf{u}, s)\mathrm{d}s\right] + V(x(t+h), t+h) \\
&\le E\left[\int_t^{t+h} f(\mathbf{X}, \mathbf{u}, s)\mathrm{d}s\right] + V(x(t+h), t+h)
\end{aligned}$$

that is:

$$V(x(t+h),t+h) - V(x,t) + E\left[\int_{t}^{t+h} f(\mathbf{X},\mathbf{u},s)\mathrm{d}s\right] \geq 0 \tag{7.2.19}$$

The above equation divided by h, take limit of $h \to 0^{+}$, there is:

$$\lim_{h\to 0^{+}} \frac{1}{h} E\left[\int_{t}^{t+h} f(\mathbf{X},\mathbf{u},s)\mathrm{d}s\right] = f(\mathbf{x},\mathbf{u},t) \tag{7.2.20}$$

Assuming that $V(\boldsymbol{x},t)$ is first-order differentiable for t and second-order differentiable for $\boldsymbol{x}$, we note that the value function is a conditional expectation like (7.2.16), and the expression within the expectation's parentheses is a function of $\boldsymbol{X}$ and t, in which $\boldsymbol{X}$ satisfies the Itô equation (7.2.1). By applying the Itô differential formula, we can derive:

$$\begin{aligned}
&\lim_{h\to 0^{+}} \frac{1}{h}[V(\mathbf{x}(t+h),t+h) - V(\mathbf{x},t)] \\
&= \frac{1}{\mathrm{d}t}\lim_{h\to 0^{+}} \int_{t}^{t+h} \left\{\frac{\partial V(\mathbf{x},s)}{\partial t} + L_x V(\mathbf{x},s)\right\}\mathrm{d}s \\
&= \frac{\partial V}{\partial t} + L_x V = \frac{\mathrm{d}V(\mathbf{x},t)}{\mathrm{d}t}
\end{aligned} \tag{7.2.21}$$

In the equation

$$L_x V = \frac{1}{2}\sigma_{il}\sigma_{jl}\frac{\partial^2 V}{\partial x_i \partial x_j} + m_i \frac{\partial V}{\partial x_i} \tag{7.2.22}$$

By doing (7.2.19)–(7.2.21), we can obtain:

$$\frac{\partial V}{\partial t} + L_x V + f(\mathbf{x},\mathbf{u},t) \geq 0 \tag{7.2.23}$$

On the other hand, if the optimal control $\boldsymbol{u}^{*}$ is also taken over the time interval $[t, t+h]$, there is:

$$\frac{\partial V}{\partial t} + L_x V + f(\mathbf{x},\mathbf{u}^{*},t) = 0 \tag{7.2.24}$$

By combining (7.2.23) and (7.2.24), we can obtain:

$$\begin{aligned}
\frac{\partial V}{\partial t} &= -\inf_{\mathbf{u}\in U}[L_x V + f(\mathbf{x},\mathbf{u},t)] \\
&= L_x^{u^{*}} V + f(\mathbf{x},\mathbf{u}^{*},t)
\end{aligned} \tag{7.2.25}$$

In the equation, $L_x^{u^*}$ is the L_x when $u = u^*$. Equation (7.2.25) must meet the final condition as per (7.2.17).

$$V(\mathbf{x}, t_f) = g(\mathbf{x}(t_f)) \tag{7.2.26}$$

Equation (7.2.25) is also often written as:

$$-\frac{\partial V}{\partial t} + \sup_{\mathbf{u} \in U} G(-V_{xx}, -V_x, \mathbf{x}, \mathbf{u}, t) = 0 \tag{7.2.27}$$

In the equation

$$G(-V_{xx}, -V_x, \mathbf{x}, \mathbf{u}, t) = -L_x V - f(\mathbf{x}, \mathbf{u}, t) \tag{7.2.28}$$

sup means least upper bound, and can be replaced by max when the value can really be achieved. Equation (7.2.19) is a finite form stochastic dynamic programming equation, Eq. (7.2.25) or (7.2.27) is a differential form stochastic dynamic programming equation, and (7.2.27) is often called the Hamilton-Jacobi-Bellman equation, or HJB equation for short. In (7.2.28), G is called general Hamiltonian function. The stochastic dynamic programming equation (7.2.25) or (7.2.27) is an inhomogeneous second-order nonlinear parabolic partial differential equation.

For an autonomic controlled system, $\boldsymbol{m}$, $\boldsymbol{\sigma}$ and $\boldsymbol{u}$ do not contain t, and (7.2.27) is turned into:

$$\sup_{\mathbf{u} \in U} G(-V_{xx}, -V_x, \mathbf{x}, \mathbf{u}) = 0 \tag{7.2.29}$$

It is a non-homogeneous second-order nonlinear elliptic partial differential equation.

For (7.2.1), assume $\boldsymbol{m}$ and $\boldsymbol{\sigma}$ are the periodic or almost periodic functions of t, and there is a unique steady-state solution for the periodic or almost periodic function in (7.2.1). Consider traversal control (7.2.1), (7.2.2) and (7.2.7), and turn the value function

$$V(\mathbf{x}, t) = \inf_{\mathbf{u} \in U} E\left[\int_t^{t_f} (f(\mathbf{X}, \mathbf{u}, t) - J(\mathbf{u}))\mathrm{d}t\right] \tag{7.2.30}$$

The dynamic programming equation is similarly deduced by the principle of dynamic programming

$$\begin{aligned} \frac{\partial V}{\partial t} &= -\inf_{u \in U}[L_x V + f(\mathbf{x}, \mathbf{u}, t) - J(\mathbf{u})] \\ &= -L_x^{\mathbf{u}^*} V - f(\mathbf{x}, \mathbf{u}^*, t) + \gamma \end{aligned} \tag{7.2.31}$$

In the equation

$$\gamma = \lim_{t_f \to \infty} \frac{1}{t_f} \int\limits_0^{t_f} f(\mathbf{x}, \mathbf{u}^*, t)\mathrm{d}t \tag{7.2.32}$$

is the optimal average cost. When (7.2.1) is an autonomous system, $\boldsymbol{m}$, $\boldsymbol{\sigma}$ and $\boldsymbol{f}$ do not contain t, and V does not contain t as well, the dynamic programming equation (7.2.31) is turned into:

$$\inf_{\mathbf{u}\in U}[L_x V + f(\mathbf{x}, \mathbf{u})] = L_x^{u^*} V + f(\mathbf{x}, \mathbf{u}^*) = \gamma \tag{7.2.33}$$

Let Ω be the security region of (7.2.1), and $q(\boldsymbol{y},s|\boldsymbol{x},t)$ be the conditional probability density of transition (the transition probability density of a sample function that remains in Ω from t to s), the conditional reliability function is:

$$R(s|\mathbf{x}, t) = \int\limits_{\Omega} q(\mathbf{y}, s|\mathbf{x}, t)\mathrm{d}\mathbf{y} \tag{7.2.34}$$

For the reliability maximization control problem of system (7.2.1) under the control constraint (7.2.2), the value function can be obtained by the case where let $f = 0$, $g = 1$ and inf → sup for (7.2.15), and let $p(\boldsymbol{y},s|\boldsymbol{x},t)$ replaced by $q(\boldsymbol{y},s|\boldsymbol{x},t)$ for (7.2.16), that is:

$$V(\mathbf{x}, t) = \sup_{\mathbf{u}\in U} \int\limits_{\Omega} q(\mathbf{y}, s|\mathbf{x}, t)\mathrm{d}\mathbf{y}, t \leq s \leq t_f \tag{7.2.35}$$

By using a similar derivation, we can obtain the following dynamic programming equation:

$$\frac{\partial V}{\partial t} = -\sup_{\mathbf{u}\in U} L_{\mathbf{x}} V, \mathbf{x} \in \Omega \tag{7.2.36}$$

The final condition is:

$$V(\mathbf{x}, t_f) = 1 \tag{7.2.37}$$

The boundary condition is:

$$V(\mathbf{x}, t) = 0, \mathbf{x} \in \Gamma \tag{7.2.38}$$

In the equation, Γ is the boundary of security region Ω.

Let (7.2.1) be the autonomous system, and for the optimal control problem with the longest average first passage time under the control constraint (7.2.2), the value function is:

$$V(\mathbf{x}) = \sup_{u \in U} E\left[\int_{t}^{t+\tau} \mathrm{d}s\right], \mathbf{x} \in \Omega \tag{7.2.39}$$

Here τ is the first passage time, and $E[\cdot]$ should also be understood as the conditional expectation (i.e. a weighted average upon $q(\mathbf{y},s|\mathbf{x},t)$). By using a similar derivation, we can obtain the following dynamic programming equation:

$$\sup_{\mathbf{u} \in U} L_x V = -1, \mathbf{x} \in \Omega \tag{7.2.40}$$

The boundary condition is:

$$V(\mathbf{x}) = 0, \mathbf{x} \in \Gamma \tag{7.2.41}$$

(3) **Solution to Stochastic Dynamic Programming Equation**

The following steps should be taken after establishing the stochastic dynamic programming equation [4]:

(1) Determine the expression of $\boldsymbol{u}^*$ (optimal control law) as a function of V by solving inf or sup for the dynamic programming equation;
(2) Substitute $\boldsymbol{u}^*$ into the dynamic programming equation to achieve the final dynamic programming equation, and solve the latter equation under the appropriate final and boundary conditions to obtain a value function V, and substitute it into the expression $\boldsymbol{u}^*$ for optimal control;
(3) Substitute the optimal control into (7.2.1), and solve the equation or proper FPK equation for optimal state trajectory or probability density.

There are two kinds of solutions to the dynamic programming equation, one is the smooth solution, called the classical solution, the other is the continuous but non-smooth solution, called the viscous solution [4]. For the dynamic programming equations (7.2.25), (7.2.27), (7.2.31) and (7.2.36), for any vector $\boldsymbol{\alpha} \in R^n$, there exists a constant $c > 0$ which satisfies the condition:

$$b_{ij}(\mathbf{x}, \mathbf{u}, t)\alpha_i\alpha_j \geq c|\boldsymbol{\alpha}|^2 \tag{7.2.42}$$

In the equation, $b_{ij} = \sigma_{il}\sigma_{jl}$, that is, quadratic $b_{ij}\alpha_i\alpha_j$ is positive definite, then they are called uniformly parabolic, otherwise, they are degenerate parabolic. For the dynamic programming equations (7.2.29), (7.2.33) and (7.2.40), if they meet the condition

$$b_{ij}(\mathbf{x}, \mathbf{u})\alpha_i\alpha_j \geq c|\boldsymbol{\alpha}|^2 \tag{7.2.43}$$

Then they are called uniformly elliptical, otherwise known as degenerate elliptic. For uniformly parabolic and uniformly elliptic dynamic programming equations, there exists a unique classical solution under proper conditions. For degenerate parabolic and elliptic dynamic programming equations, only viscous solution can be obtained.

A continuous function $V(\boldsymbol{x},t)$ is called the viscous sub-solution of the stochastic dynamic programming equation (7.2.27) under the final condition (7.2.26) if it satisfies the condition

$$V\left(\mathbf{x}, t_f\right) \leq g\left(\mathbf{x}\left(t_f\right)\right) \tag{7.2.44}$$

And for any function $\varphi(\boldsymbol{x},t)$ that can be continuously first-order differentiable for t and continuously second-order differentiable for $\boldsymbol{x}$, once $V-\varphi$ reaches a local maximum at a certain point, there is:

$$-\frac{\partial \varphi}{\partial t} + \sup_{\mathbf{u} \in U} G(-\varphi_{xx}, -\varphi_x, \mathbf{x}, \mathbf{u}, t) \leq 0 \tag{7.2.45}$$

If the inequalities in (7.2.44) and (7.2.45) are replaced by "$\geq$", "local maximum" replaced by "local minimum," then $V(\boldsymbol{x},t)$ is viscous super-solution to (7.2.27) under the final condition (7.2.26). If $V(\boldsymbol{x},t)$ serves as viscous sub-solution and viscous super-solution at the same time, it is called a viscous solution. Viscous solutions can also be equivalently defined by substituting differentials in stochastic dynamic programming equations with so-called sub-differentials and super-differentials. If the value function is continuously first-order differentiable for t and second-order differentiable for $\boldsymbol{x}$, it can be a viscous solution only if it is a classical solution to stochastic dynamic programming. It can be seen that viscous solution is a general solution to a stochastic dynamic programming equation, and it is also unique under proper conditions.

For the stochastic optimal control problems of mechanical and structural systems, the conditions (7.2.42) and (7.2.43) are often not satisfied, so the corresponding stochastic dynamic programming equations only have viscous solutions. However, after applying the stochastic averaging method, for the stochastic control of the average Itô equation, these conditions are often satisfied, so that there are classical solutions. Therefore, combining the stochastic averaging method with the principle of dynamic programming can regularize the solutions to stochastic dynamic programming equations.

7.3 Stochastic Optimal Control Based on Power Adjustment

7.3.1 *Stochastic Optimal Control of Quasi Hamiltonian System*

A theoretical method for stochastic optimal control of the following quasi Hamiltonian system is developed for the stochastic averaging method and stochastic dynamic programming method based on quasi Hamiltonian system.

Taking a controlled quasi Hamiltonian system into consideration, the motion equation is [4, 11]

$$\begin{aligned} \dot{Q}_i &= \frac{\partial H'}{\partial P_i} \\ \dot{P}_i &= -\frac{\partial H'}{\partial Q_i} - \varepsilon c'_{ij}(\mathbf{Q},\mathbf{P})\frac{\partial H'}{\partial P_j} + u_i(\mathbf{Q},\mathbf{P}) + \varepsilon^{1/2} f'_{ik}(\mathbf{Q},\mathbf{P})\xi_k(t) \\ i,j &= 1,2,\ldots,n;\ k = 1,2,\ldots,m \end{aligned} \tag{7.3.1}$$

In the equation, $H' = H'(\mathbf{Q},\mathbf{P})$ and $\varepsilon c'_{ij}$ are respectively Hamiltonian function and quasi-linear damping coefficient of an uncontrolled and non-excited system; $\xi_k(t)$ is a stochastic process which can include harmonic function; and $\varepsilon^{1/2} f'_{ik}(\mathbf{Q},\mathbf{P})$ is an amplitude of disturbances. When $\xi_k(t)$ is white noise and the correlation function is $2D_{\mathrm{kl}}\delta(\tau)$, Equation (7.3.1) can be modeled as Itô stochastic differential equation as follows:

$$\begin{aligned} \mathrm{d}Q_i &= \frac{\partial H''}{\partial P_i}\mathrm{d}t \\ \mathrm{d}P_i &= -\left(\frac{\partial H''}{\partial Q_i} - \varepsilon m'_{ij}(\mathbf{Q},\mathbf{P})\frac{\partial H''}{\partial P_j} - u_i(\mathbf{Q},\mathbf{P})\right)\mathrm{d}t + \varepsilon^{1/2}\boldsymbol{\sigma}'_{ik}(\mathbf{Q},\mathbf{P})\mathrm{d}B_k(t) \\ i,j &= 1,2,\ldots,n;\ k = 1,2,\ldots,m \end{aligned} \tag{7.3.2}$$

In the equation, $H'' = H''(\mathbf{Q},\mathbf{P})$ and $\varepsilon m'_{ij}$ are respectively the Hamiltonian function and quasi-linear damping coefficient modified by the Wong-Zakai correction; $\boldsymbol{\sigma}'\boldsymbol{\sigma}'^{T} = 2\mathbf{f}'\mathbf{D}\mathbf{f}'^{T}$. In (7.3.1) and (7.3.2), u_{i} is general feedback control force. If there are β control actuators in the system, the control force given by each actuator is $U_\alpha = U_\alpha(\boldsymbol{Q},\boldsymbol{P})$, then $u_{\mathrm{i}} = g_{\mathrm{ia}}U_\alpha$, $[g_{\mathrm{ia}}]$ is the matrix for the control actuator placement. If g_{ia} is independent of $\boldsymbol{Q}$ and $\boldsymbol{P}$, then call the corresponding u_{i} the additional control force. If g_{ia} is the function of $\boldsymbol{Q}$ and $\boldsymbol{P}$, then u_{i} is the parameter control force.

The purpose of the study is to find the optimal feedback control law to minimize the system response and predict the optimal control system response. The basic idea of stochastic optimal control strategy is to divide the feedback control force u_i into conservative control force $u_i^{(1)}$ and dissipative control force $u_i^{(2)}$, and change the Hamiltonian structure of system with $u_i^{(1)}$ so as to change the distribution of energy and response in the system. Use $u_i^{(2)}$ to dissipate system energy, increasing stability and reducing system energy and response. $u_i^{(1)}$ is determined by conservative Hamiltonian system optimal control theory, $u_i^{(2)}$ is determined by stochastic dynamic programming based on the stochastic average of the quasi Hamiltonian system [4].

The conservative Hamiltonian system or the conservative Lagrange system, that is, the uncontrolled and controlled systems are either Hamiltonian systems or Lagrange systems, which is a complex problem in the control of mechanical systems [4]. So far, the main research is on how to change the kinetic or potential energy to make uncontrolled Hamiltonian systems tend to be stable. And with conservative control $u_i^{(1)}$, the integrability and resonance of the system can be changed, so that the distribution of energy and response within the system is as satisfying as possible. In principle, this can be referred to the following optimal control problem: let H' denote the Hamiltonian function of the uncontrolled system, H be the required Hamiltonian function, and select $u_i^{(1)}$ to minimize a certain performance index, that is:

$$\inf_{\mathbf{u}^{(1)}} J\left(H', H, \mathbf{u}^{(1)}\right) \tag{7.3.3}$$

In the equation, $\mathbf{u}^{(1)} = \left[u_1^{(1)} \quad u_2^{(1)} \quad \cdots \quad u_n^{(1)}\right]^T$. There is currently no general solution to this problem. The following is only a description of specific examples.

After determining $u_i^{(1)}$, combining $u_i^{(1)}$ with $-\partial H'/\partial Q_i$ or $-\partial H''/\partial Q_i$ to get an optimal Hamiltonian function $H = H(\boldsymbol{Q},\boldsymbol{P})$. Equations (7.3.1) and (7.3.2) respectively turn into:

$$\begin{aligned}
\dot{Q}_i &= \frac{\partial H}{\partial P_i} \\
\dot{P}_i &= -\frac{\partial H}{\partial Q_i} - \varepsilon c'_{ij}(\mathbf{Q},\mathbf{P})\frac{\partial H}{\partial P_j} + u_i^{(2)}(\mathbf{Q},\mathbf{P}) + \varepsilon^{1/2} f'_{ik}(\mathbf{Q},\mathbf{P})\xi_k(t) \\
i,j &= 1,2,\ldots,n;\ k = 1,2,\ldots,m
\end{aligned} \tag{7.3.4}$$

and

$$\begin{aligned}
\mathrm{d}Q_i &= \frac{\partial H}{\partial P_i}\mathrm{d}t \\
\mathrm{d}P_i &= -\left(\frac{\partial H}{\partial Q_i} - \varepsilon m'_{ij}(\mathbf{Q},\mathbf{P})\frac{\partial H}{\partial P_j} - u_i^{(2)}(\mathbf{Q},\mathbf{P})\right)\mathrm{d}t + \varepsilon^{1/2}\boldsymbol{\sigma}'_{ik}(\mathbf{Q},\mathbf{P})\mathrm{d}B_k(t) \\
i,j &= 1,2,\ldots,n;\ k = 1,2,\ldots,m
\end{aligned} \tag{7.3.5}$$

Although direct application of the stochastic dynamic programming method to (7.3.5), and for (7.3.4), after introducing a linear or non-linear filter for disturbance $\xi_k(t)$ to make it an augmented Itô stochastic differential equation, the method of stochastic dynamic programming can also be applied, the dynamic programming equation is then $2n$-dimensional or higher dimensional, and the diffusion matrix is often degenerate and can not satisfy the condition (7.2.42) or (7.2.43), so that there is no classical solution. One of the methods to avoid the above-mentioned situation is to apply the stochastic averaging method of the quasi Hamiltonian system to (7.3.4) or (7.3.5), and apply the stochastic dynamic programming method to the average Itô equation. In this way, not only can the dimension of the dynamic programming equation be reduced, but also the diffusion matrix becomes non-degenerate, that is, the condition (7.2.42) or (7.2.43) is satisfied, thus making the dynamic programming equation have a classical solution [4].

To apply the stochastic averaging method of the quasi Hamiltonian system, it is needed to assume that $u_i^{(2)}$ in (7.3.4) or (7.3.5) is the ε-order small quantity. The corresponding control is called weak control. In the application, as long as the energy difference between the energy of the stochastic disturbances input system and the energy consumed by damping force and the control force is smaller compared with the energy of the system itself, the stochastic averaging method of the quasi Hamiltonian system can be applied [4].

Apply the stochastic averaging method of the quasi non-integrable Hamiltonian system to (7.3.5), we can obtain an Itô equation:

$$\mathrm{d}H = \left[\bar{m}(H) + \left\langle u_i^{(2)} \frac{\partial H}{\partial P_i} \right\rangle\right]\mathrm{d}t + \bar{\sigma}(H)\mathrm{d}B(t) \tag{7.3.6}$$

In the equation, $\bar{m}(H)$ and $\bar{\sigma}(H)$ are determined according to the stochastic averaging method of the quasi Hamiltonian system, and

$$\langle\cdot\rangle = \frac{1}{T(H)} \int_{\Omega} \left(\cdot / \frac{\partial H}{\partial p_1}\right) \mathrm{d}q_1 \ldots \mathrm{d}q_n \mathrm{d}p_2 \ldots \mathrm{d}p_n \tag{7.3.7}$$

It should be noted that $u_i^{(2)}$ is a function of $\boldsymbol{Q}$ and $\boldsymbol{P}$, and $u_i^{(2)} \partial H / \partial P_i$ is a function of H. Since $u_i^{(2)}$ has not yet been determined, the average of the second term on the right side of (7.3.6) can not be completed for now.

Consider control of (7.3.6) over a finite time interval $[0, t_f]$. Assume the performance index as:

$$J\left(\mathbf{u}^{(2)}\right) = E\left[\int_0^{t_f} f\left(H(s), \left\langle \mathbf{u}^{(2)}(s) \right\rangle\right) \mathrm{d}s + g\left(H\left(t_f\right)\right)\right] \tag{7.3.8}$$

Introduce the value function:

$$V(H,t)=\inf_{u^{(2)}}E\left[\int_t^{t_f}f\left(H(s),\left\langle\mathbf{u}^{(2)}(s)\right\rangle\right)\mathrm{d}s+g(H(t_f))\right] \tag{7.3.9}$$

Similar to (7.2.19)–(7.2.26), the following dynamic programming equations can be established for the control problems (7.3.6) and (7.3.8):

$$\frac{\partial V}{\partial t}=-\inf_{\mathbf{u}^{(2)}}\left\{\frac{1}{2}\bar{\sigma}^2(H)\frac{\partial^2 V}{\partial H^2}+\left[\bar{m}(H)+\left\langle u_i^{(2)}\frac{\partial H}{\partial p_i}\right\rangle\right]\frac{\partial V}{\partial H}+f\left(H,\left\langle\mathbf{u}^{(2)}\right\rangle\right)\right\} \tag{7.3.10}$$

And the final condition is:

$$V(H,t_f)=g(H(t_f)) \tag{7.3.11}$$

The necessary condition for the right side of (7.3.10) to take the minimal value is:

$$\frac{\partial}{\partial u_i^{(2)}}\left[\left\langle u_i^{(2)}\frac{\partial H}{\partial p_i}\right\rangle\frac{\partial V}{\partial H}+f\left(H,\left\langle\mathbf{u}^{(2)}\right\rangle\right)\right]=0, i=1,2,\ldots,n \tag{7.3.12}$$

Therefore, we can determine the optimal control law $u_i^{(2)*}$.

Substitute $u_i^{(2)*}$ as determined by the above procedures into (7.3.10) to replace $u_i^{(2)}$, and complete the averaging operation as per (7.3.7) to obtain the final dynamic programming equation.

In the above derivation, the size of the control force is not limited, and the best unbounded control is obtained in turn. If the control force is bounded, that is:

$$|u_i|\le b_i, b_i>0,\ i=1,2,\ldots,n \tag{7.3.13}$$

Then the corresponding optimal control is called optimal bounded control. At this point, the performance index has nothing to do with $u_i^{(2)}$, and the form is:

$$J_1=E\left[\int_0^{t_f}f_1(H(s))\mathrm{d}s+g(H(t_f))\right] \tag{7.3.14}$$

The value function is defined as:

$$V_1(H,t)=\inf_{u^{(2)}}E\left[\int_t^{t_f}f_1(H(s))\mathrm{d}s+g(H(t_f))\right] \tag{7.3.15}$$

The corresponding dynamic programming equation is:

$$\frac{\partial V_1}{\partial t} = -\inf_{\mathbf{u}^{(2)}}\left\{\frac{1}{2}\bar{\sigma}^2(H)\frac{\partial^2 V_1}{\partial H^2} + \left[\bar{m}(H) + \left\langle u_i^{(2)}\frac{\partial H}{\partial p_i}\right\rangle\right]\frac{\partial V_1}{\partial H} + f_1(H)\right\} \tag{7.3.16}$$

The final condition is:

$$V_1(H, t_f) = g(H(t_f)) \tag{7.3.17}$$

To solve the minimal value for $u_i^{(2)}$ by the right side of (7.3.16), taking into account of (7.3.13), we can obtain:

$$u_i^{(2)*} = -b_i \text{sgn}\left(\frac{\partial H}{\partial P_i}\frac{\partial V_1}{\partial H}\right) \tag{7.3.18}$$

By (7.3.15), the positive/negative value of $\partial V_1/\partial H$ depends on $f_1'(H)$, and if it takes $f_1'(H) > 0$, then

$$u_i^{(2)*} = -b_i \text{sgn}\left(\frac{\partial H}{\partial P_i}\right) = u_i^{(2)*} = -b_i \text{sgn}(\dot{Q}_i) \tag{7.3.19}$$

It is dry friction-type control or bang-bang control, the size of the control force unchanged, and in the opposite direction from the general speed. It changes its direction at $\dot{Q} = 0$. Because $u_i^{(2)*}$ has nothing to do with V_1, it is unnecessary to solve the dynamic programming equation.

Finally, substitute the optimal control force $u_i^{(2)*}$ in the (7.3.19) into average Itô equation (7.3.6) to achieve an average of $u_i^{(2)*}\partial H/\partial P_i$ so as to obtain an average Itô equation for the optimal control system:

$$\mathrm{d}H = \bar{\bar{m}}(H)\mathrm{d}t + \bar{\sigma}(H)\mathrm{d}B(t) \tag{7.3.20}$$

In the equation

$$\bar{\bar{m}}(H) = \bar{m}(H) + \left\langle u_i^{(2)*}\frac{\partial H}{\partial P_i}\right\rangle \tag{7.3.21}$$

By solving the average FPK equation corresponding to the average Itô equation (7.3.20), we can obtain the response statistics of the optimal control system.

To evaluate an optimal control strategy, the following two criteria are introduced. One is the control effect [4, 8]:

$$k_h = \frac{\sigma_h^u - \sigma_h^c}{\sigma_h^u} \tag{7.3.22}$$

In the equation, σ is the standard deviation; $h = h(\boldsymbol{Q}, \boldsymbol{P})$ is the test function such as $h = Q_i^2$;the superscript u means uncontrolled system and the superscript c means the controlled system. k denotes that the standard deviation of a certain response variable caused by the optimal control is relatively reduced. The other is the control efficiency [4, 8]:

$$\mu_h = \frac{k_h}{\sigma_{u^*}} \tag{7.3.23}$$

It indicates the control effect of unit standard deviation control force. Obviously, the greater the control effect and control efficiency, the better the control strategy.

7.3.2 Stochastic Optimal Control Strategy Based on Power Adjustment

Based on some approximate assumptions, the generators adopt the second-order model and consider the controlled multi-machine power system. Under stochastic disturbance, the electromechanical transient process model is as follows [12]:

$$\begin{cases} \frac{d\delta_i}{d_t} = \omega_N \omega_i \\ M_i \frac{d\omega_i}{d_t} = P_i - P_{ei} - D_i\omega_i + \xi_i(t) + u_i \\ i = 1, 2, \ldots, n \end{cases} \tag{7.3.24}$$

Wherein:

$$\begin{cases} P_i = P_{mi} - G_{ii}E_i^2 \\ P_{ei} = \sum_{j=1, j\neq i}^{n} E_iE_jB_{ij}\sin(\delta_i - \delta_j) \end{cases} \tag{7.3.25}$$

In the equation: ω_i, δ_i, M_i, D_i and P_{mi} are respectively the speed deviation, power angle, inertia time constant, damping coefficient and mechanical power of the ith generator; $\xi_i(t)$ is the Gaussian White noise with mean being 0 and variance being σ_i^2; u_i is the system control variable, E_i is the internal potential of the ith generator, G_{ii} is the ith diagonal element in the conductance matrix, B_{ij} is the off-diagonal element in the ith row and jth column of the susceptance. In the equation, M_i, t and δ_i are actual values, the remaining quantities are per-unit values.

You need to express ξ_i(t) by independent standard Gaussian white noise $W_i(t)$ before converting a multi-machine power system model into a Stratonovich stochastic differential equation, so the multi-machine system can be expressed as:

$$\begin{cases} \frac{d\delta_i}{d_t} = \omega_N \omega_i \\ M_i \frac{d\omega_i}{d_t} = P_i - P_{ei} - D_i\omega_i + \sigma_i W_i(t) + u_i \\ i = 1, 2, \ldots, n \end{cases} \tag{7.3.26}$$

According to the theory of stochastic dynamics, the Itô-type stochastic differential equations of the above multi-machine power system can be expressed as:

$$\begin{cases} \mathrm{d}\delta_i = \omega_{\mathrm{N}}\omega_i \mathrm{d}t \\ \mathrm{d}\omega_i = \dfrac{1}{M_i}(P_i - P_{ei} - D_i\omega_i + u_i)\mathrm{d}t + \dfrac{\sigma_i}{M_i}\mathrm{d}B(t) \\ i = 1, 2, \ldots, n \end{cases} \tag{7.3.27}$$

Physically, after one period of vibration, if the difference between the energy of the stochastic disturbance input system and the energy consumed by the damping is small compared with the energy of the system itself, it can be regarded as the quasi Hamiltonian system. The energy function of the above multi-machine power system adopts the transient energy function commonly used in the power system [13]:

$$\begin{aligned} H = {} & \frac{1}{2}\sum_{i=1}^{n} M_i\omega_N\omega_i - \sum_{i=1}^{n} P_i(\delta_i - \delta_{is}) \\ & - \left[\sum_{i=1}^{n}\sum_{i=j+1}^{n} E_iE_jB_{ij}\cos(\delta_i - \delta_j) - \sum_{i=1}^{n}\sum_{i=j+1}^{n} E_iE_jB_{ij}\cos(\delta_{is} - \delta_{js})\right] \end{aligned} \tag{7.3.28}$$

In the equation: δ_{is} and δ_{js} are the power angles of the ith generator and the jth generator respectively. $\frac{1}{2}\sum_{i=1}^{n} M_i\omega_N\omega_i$ is the variable quantity of kinetic energy of all the generators, and the rest is the variable quantity of the potential energy of all the generators.

According to (7.3.28), we can obtain:

$$\begin{cases} \dfrac{\partial H}{\partial \omega_i} = M_i\omega_{\mathrm{N}}\omega_i \\ \dfrac{\partial H}{\partial \delta_i} = -P_i + \sum\limits_{j=1, j\neq i}^{n} E_iE_jB_{ij}\sin(\delta_i - \delta_j) = -P_i + P_{\mathrm{ei}} \\ \dfrac{\partial^2 H}{\partial \omega_i^2} = M_i\omega_{\mathrm{N}} \\ i = 1, 2, \ldots, n \end{cases} \tag{7.3.29}$$

By substituting the (7.3.29) into (7.3.27), we can obtain the quasi Hamiltonian system equation for multi-machine power system:

$$\begin{cases} \mathrm{d}\delta_i = \dfrac{1}{M_i}\dfrac{\partial H}{\partial \omega_i}\mathrm{d}t \\ \mathrm{d}\omega_i = \left(-\dfrac{1}{M_i}\dfrac{\partial H}{\partial \delta_i} - \dfrac{D_i}{M_i^2\omega_{\mathrm{N}}}\dfrac{\partial H}{\partial \omega_i} + \dfrac{u_i}{M_i}\right)\mathrm{d}t + \dfrac{\sigma_i}{M_i}\mathrm{d}B_i(t) \end{cases} \tag{7.3.30}$$

If there is no stochastic disturbance term $\frac{\sigma_i}{M_i}\mathrm{d}B_i(t)$ in (7.3.30), it can be written based on the full differential equation:

$$\mathrm{d}H = \sum_{i=1}^{n}\left(\frac{\partial H}{\partial \delta_i}\mathrm{d}\delta_i + \frac{\partial H}{\partial \omega_i}\mathrm{d}\omega_i\right) \tag{7.3.31}$$

Considering the stochastic disturbance term, we need to introduce the Wong-Zakai correction term into the total differential equation and then substitute Eq. (7.3.30) into the total differential equation when deriving the Itô equation of the energy function, then we can obtain after arranging:

$$\mathrm{d}H = \sum_{i=1}^{n}\left\{\left[-\frac{D_i}{M_i^2\omega_{\mathrm{N}}}\left(\frac{\partial H}{\partial \omega_i}\right)^2 + \frac{1}{2}\frac{\sigma_i^2}{M_i^2}\frac{\partial^2 H}{\partial \omega_i^2}\right]\mathrm{d}t + \frac{u_i}{M_i}\frac{\partial H}{\partial \omega_i}\mathrm{d}t + \frac{\sigma_i}{M_i}\frac{\partial H}{\partial \omega_i}\mathrm{d}B_i(t)\right\} \tag{7.3.32}$$

The second term in (7.3.32) is the Wong-Zakai correction term, and substitute (7.3.29) into (7.3.32) to obtain the Itô equation for energy function as follows:

$$\mathrm{d}H = \sum_{i=1}^{n}\left(-D_i\omega_{\mathrm{N}}\omega_i^2 + \frac{\sigma^2\omega_{\mathrm{N}}}{2M_i} + \frac{u_i}{M_i}\frac{\partial H}{\partial \omega_i}\right)\mathrm{d}t + \sum_{i=1}^{n}\sigma_i\omega_{\mathrm{N}}\omega_i\mathrm{d}B_i(t) \tag{7.3.33}$$

If both the system damping coefficient and the stochastic disturbance intensity are small parameters and the control is weak, the system energy H weakly converges to the one-dimensional diffusion process:

$$\mathrm{d}H = \left[\bar{m}(H) + \sum_{i=1}^{n}\left\langle\frac{u_i}{M_i}\frac{\partial H}{\partial \omega_i}\right\rangle\right]\mathrm{d}t + \bar{\sigma}(H)\mathrm{d}B(t) \tag{7.3.34}$$

In the equation:

$$\bar{m}(H) = \frac{1}{T(H)}\int_{\Omega}\left[\sum_{i=1}^{n}\left(-D_i\omega_{\mathrm{N}}\omega_i^2 + \frac{\sigma_i^2\omega_{\mathrm{N}}}{2M_i}\right)\Big/\frac{\partial H}{\partial \omega_1}\right]\mathrm{d}\delta_1\ldots\mathrm{d}\delta_n\mathrm{d}\omega_2\ldots\mathrm{d}\omega_n$$

$$\bar{\sigma}^2(H) = \frac{1}{T(H)}\int_{\Omega}\left[\sum_{i=1}^{n}(\sigma_i\omega_N\omega_i)^2\Big/\frac{\partial H}{\partial \omega_1}\right]\mathrm{d}\delta_1\ldots\mathrm{d}\delta_n\mathrm{d}\omega_2\ldots\mathrm{d}\omega_n$$

$$\left\langle\frac{u_i}{M_i}\frac{\partial H}{\partial \omega_i}\right\rangle = \frac{1}{T(H)}\int_{\Omega}\left[\left(\frac{u_i}{M_i}\frac{\partial H}{\partial \omega_i}\right)\Big/\frac{\partial H}{\partial \omega_1}\right]\mathrm{d}\delta_1\ldots\mathrm{d}\delta_n\mathrm{d}\omega_2\ldots\mathrm{d}\omega_n$$

$$T(H) = \int_{\Omega} \left(1/\frac{\partial H}{\partial \omega_1}\right) \mathrm{d}\delta_1 \ldots \mathrm{d}\delta_n \mathrm{d}\omega_2 \ldots \mathrm{d}\omega_n$$

$$\Omega = \{(\delta_1, \ldots, \delta_n, \omega_2, \ldots, \omega_n) | H(\delta_1, \ldots, \delta_n, 0, \omega_2, \ldots, \omega_n) \leq H\}$$

The stochastic dynamic programming method is to establish and solve the stochastic dynamic programming equation for a given stochastic optimal control problem, determine the optimal control, and then solve the optimal state.

Assume the energy function $H(t)$ can be stochastically changed on $[0, \infty]$, and the system bounded fluctuation region is $[0, H_c]$, the initial value is $H(0) = H_0 \in [0, H_c)$, and take the probability of that the system stays in the bounded fluctuation region D as the performance index, i.e. the reliability function:

$$J(u) = P(H(t) \in D, t_0 \leq t \leq \tau) \tag{7.3.35}$$

The problem of nonlinear stochastic optimal control with the goal of maximum reliability is studied. Define the value function as:

$$V(H, t) = \sup_{u \in U} P\{H(s, u) \in [0, H_c), t < s \leq t_f | H(t, u) \in [0, H_c)\} \tag{7.3.36}$$

In the equation: $u \in U$ is control constraint, $t_f \leq \tau$ is final control time, and τ is the first passage time.

According to the above value function, the dynamic programming equation of the stochastic optimal control problem with the maximum reliability as the goal is:

$$\begin{gathered} \frac{\partial V}{\partial t} = -\sup_{u \in U}\left\{\frac{1}{2}\bar{\sigma}^2(H)\frac{\partial^2 V}{\partial H^2} + \left[\bar{m}(H) + \sum_{i=1}^{n}\left\langle \frac{u_i}{M_i}\frac{\partial H}{\partial \omega_i}\right\rangle\right]\frac{\partial V}{\partial H}\right\} \\ 0 \leq t \leq t_f, H \in [0, H_c) \end{gathered} \tag{7.3.37}$$

The boundary conditions are:

$$\begin{cases} V(H_c, t) = 0 \\ V(0, t) = \text{limited} \end{cases} \tag{7.3.38}$$

The final condition is:

$$V(H, t_f) = 1, H \in [0, H_c) \tag{7.3.39}$$

Let the control constraint be $\left|\frac{u_i}{M_i}\right| \leq b_i$ and b_i be a positive constant. Obviously, when $\left|\frac{u_i}{M_i}\right| = b_i$ and the u_i makes $\frac{u_i}{M_i}\frac{\partial H}{\partial \omega_i}\frac{\partial V}{\partial H}$ positive, the right side of (7.3.37) takes a maximal value. Therefore, the optimal control law is:

$$\frac{u_i^*}{M_i} = b_i \mathrm{sgn}\left(\frac{\partial H}{\partial \omega_i}\frac{\partial V}{\partial H}\right) \tag{7.3.40}$$

According to the knowledge of probability theory, the reliability function is a monotonically decreasing function, thus $\frac{\partial V}{\partial H} < 0$, and (7.3.40) can be simplified as:

$$\frac{u_i^*}{M_i} = -b_i \mathrm{sgn}\left(\frac{\partial H}{\partial \omega_i}\right) \tag{7.3.41}$$

Substituting the above optimal control law into the (7.3.34) to complete the average to obtain the average Itô equation of the quasi non-integrable Hamiltonian system under optimal control as follows:

$$\mathrm{d}H = \bar{\bar{m}}(H)\mathrm{d}t + \bar{\sigma}(H)\mathrm{d}B(t) \tag{7.3.42}$$

In the equation:

$$\bar{\bar{m}}(H) = \bar{m}(H) + \frac{1}{T(H)}\int_{\Omega}\left(\sum_{i=1}^{n}\left(\frac{u_i^*}{M_i}\frac{\partial H}{\partial \omega_i}\right)\Big/\frac{\partial H}{\partial \omega_1}\right)\mathrm{d}\delta_1\ldots\mathrm{d}\delta_n\mathrm{d}\omega_2\ldots\mathrm{d}\omega_n$$

Taking the stochastic disturbances into account, the system state becomes a stochastic vector, so whether the system state quantity can be kept in the region is a stochastic event and needs to be described by the probability, that is, the intra-region probability.

According to the quasi non-integrable Hamiltonian system first passage time theory, the conditional reliability function for the above system is:

$$R^c(t_1|H_0) = P\{\, H(s,u^*) \in [0,H_c), s \in (0,t_1]|H(0) \in [0,H_c)\} \tag{7.3.43}$$

The conditional reliability function satisfies the backward Kolmogorov equation as follows:

$$\frac{\partial R^c}{\partial t_1} = \bar{\bar{m}}(H_0)\frac{\partial R^c}{\partial H_0} + \frac{1}{2}\bar{\sigma}^2(H_0)\frac{\partial^2 R^c}{\partial H_0^2} \tag{7.3.44}$$

The boundary conditions are:

$$\begin{cases} R^c(t_1|H_c) = 0 \\ R^c(t_1|0) = \text{limited} \end{cases} \tag{7.3.45}$$

The initial condition is:

$$R^c(0|H_0) = 1,\ H_0 \in [0,H_c) \tag{7.3.46}$$

By solving (7.3.44)–(7.3.46), we can obtain the conditional reliability function for optimal controllable quasi non-integrable Hamiltonian system.

7.3.3 Simulation and Analysis of Control Effect

The four-machine two-area system is selected as shown in Fig. 7.1 [14].

Select parameters for the four-machine two-area system as follows:

$\omega_N = 2*\pi*60$; $M_1 = 13$ s; $M_2 = 13$ s; $M_3 = 12.35$ s; $M_4 = 12.35$ s; $E_1 = 1.03$; $E_2 = 1.01$; $E_3 = 1.03$; $E_4 = 1.01$; $P_1 = 0.9981$; $P_2 = 0.3198$; $P_3 = -0.2415$; $P_4 = -1.0765$; $B_{12} = B_{21} = 1.49$; $B_{13} = B_{31} = 0.62$; $B_{14} = B_{41} = 0.70$; $B_{23} = B_{32} = 0.70$; $B_{24} = B_{42} = 0.76$; $B_{34} = B_{43} = 1.49$; $D_1 = D_2 = D_3 = D_4 = 3$; $\sigma_1 = \sigma_2 = \sigma_3 = \sigma_4 = 0.07$.

According to (7.3.24), the steady states of the system are $\delta_{1s} - \delta_{4s} = 37.2° = 0.649$ rad, $\delta_{2s} - \delta_{4s} = 27.5° = 0.480$ rad and $\delta_{3s} - \delta_{4s} = 10.2° = 0.178$ rad. The critical energy of the system, given by (7.3.28), is 2.17, and take 2.1 as the maximum energy of the bounded fluctuation region.

By substituting the parameters, the numerical solutions of $\bar{m}(H)$ and $\bar{\sigma}^2(H)$ without control are shown in Fig. 7.2.

Select an appropriate control constraint, calculate the optimal control law of the system, and by adding the control, the drift coefficient of the average equation will be changed, which is shown in Fig. 7.3.

As can be seen from the figure, when the control constraint increases, the drift coefficient of the system decreases and it is proved that such control is effective.

Based on the obtained drift coefficients and diffusion coefficients, the analytical solutions of the conditional reliability function are as shown in Fig. 7.4.

In Fig. 7.4, the discrete points are the Monte Carlo solution to (7.3.24), and the continuous curve is the analytical solution to the conditional reliability function.

As shown in Fig. 7.4, the Monte Carlo solution fits well with the analytical solution which proves the accuracy of the stochastic averaging method. Besides when the control constraint increases, the reliability of the system increases, which proves the effectiveness of the control.

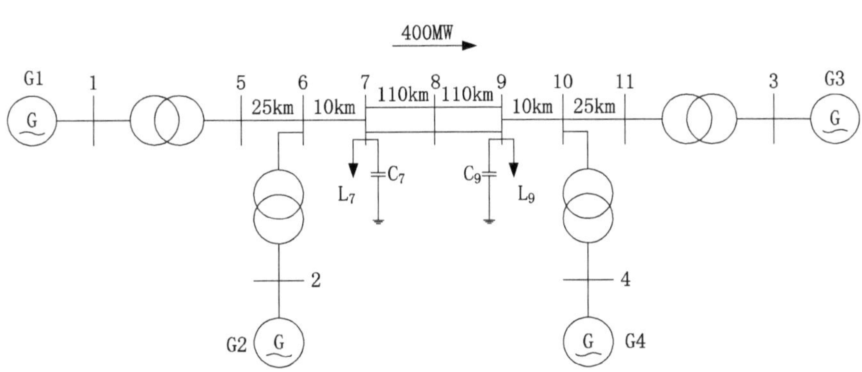

Fig. 7.1 Four-machine two-area system

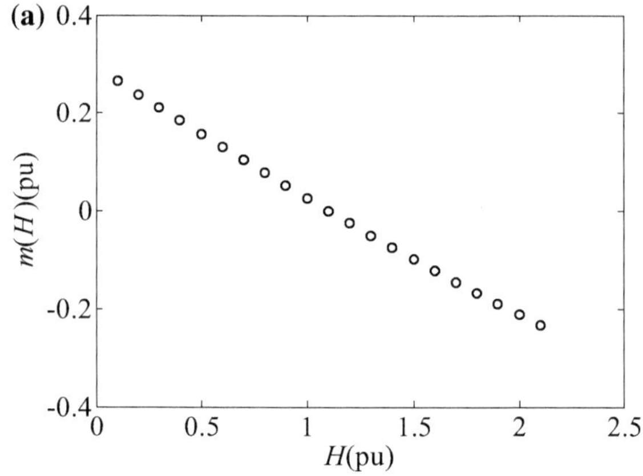

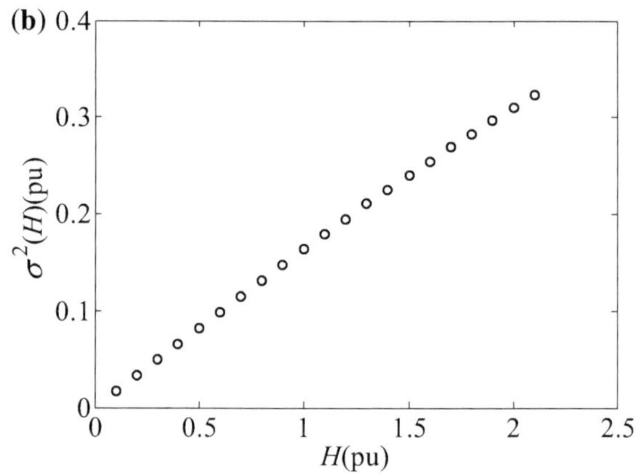

Fig. 7.2 **a** Drift coefficient and **b** diffusion coefficient

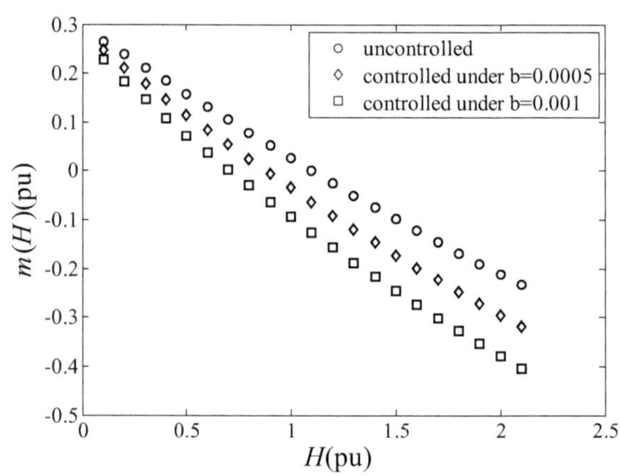

Fig. 7.3 Drift coefficients under different control constraints

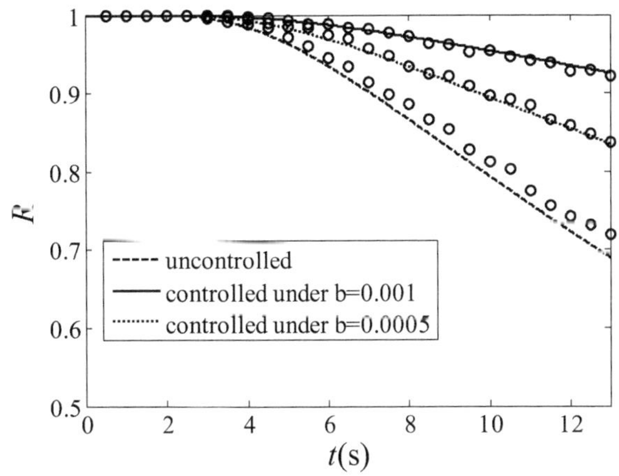

Fig. 7.4 Conditional reliability function

7.4 Stochastic Optimal Control Based on Excitation Adjustment

7.4.1 Stochastic Optimal Control of Quasi General Hamilton System

The biggest difference between the general Hamiltonian system and the traditional Hamiltonian system is that the former's dimension can be an odd number, and its structure matrix can be a function of the state variables.

Consider the following quasi general Hamiltonian system [7]:

$$\dot{X}_i = \left[X_i, H'\right] + \varepsilon d'_{ij}(\mathbf{X})\frac{\partial H'}{\partial X_j} + \varepsilon^{1/2} f_{il}(\mathbf{X}) W_l(t) \\ i,j = 1,\ldots,m;\ l = 1,\ldots,N \tag{7.4.1}$$

Wherein $X = \{X_1, \ldots, X_m\}^T$, while $H' = H'(\mathbf{X})$ is the general Hamiltonian function. $\left[X_i, H'\right] = \sum_{i,j=1}^{m} J_{ij}(\mathbf{X}) \frac{\partial X_i}{\partial X_i}\frac{\partial H'}{\partial X_j}$ is the Poisson bracket of X_i and H', $J_{ij}(\mathbf{X})$ is the structural matrix of Poisson brackets. $\varepsilon d'_{ij}(\mathbf{X})$ is the quasi-linear damping coefficient, $\varepsilon^{1/2} f_{il}(\mathbf{X})$ is the amplitude of stochastic disturbance, and $W_l(t)$ is Gaussian white noise, of which correlation function is $[W_{l_1}(t) W_{l_2}(t+\tau)] = 2D_{l_1 l_2}\delta(\tau)(l_1, l_2 = 1,\ldots,N)$.

Equation (7.4.1) can be transformed into the following Itô stochastic differential equation:

$$\begin{aligned} dX_i &= \left\{ \left[X_i, H'\right] + \varepsilon d'_{ij}(\mathbf{X}) \frac{\partial H'}{\partial X_j} + D_{l_1 l_2} f_{jl_2} \frac{\partial f_{il_1}}{\partial X_j} \right\} dt + \varepsilon^{1/2} \sigma_{il} dB_l(t) \\ i,j &= 1, \ldots, m;\ l, l_1, l_2 = 1, \ldots, N \end{aligned} \tag{7.4.2}$$

Wherein $B_l(t)$ is the standard Wiener process, that is, $\boldsymbol{\sigma\sigma}^T = 2\mathbf{fDf}^T$.

The Wong-Zakai correction term $D_{l_1 l_2} f_{jl_2} \frac{\partial f_{il_1}}{\partial X_j}$ in (7.4.2) can be divided into the conservative part and the dissipative part. By combining the conservative part with $\left[X_i, H'\right]$, a new Hamiltonian function H can be constructed. Combining the dissipative part with the damping term makes a new damping coefficient $\varepsilon d_{ij}(\mathbf{X})$ as well. Therefore Eq. (7.4.2) can be rewritten as:

$$\begin{aligned} dX_i &= \left\{ [X_i, H] + \varepsilon d_{ij} \frac{\partial H}{\partial X_j} \right\} dt + \varepsilon^{1/2} \sigma_{il} dB_l(t) \\ i,j &= 1, \ldots, m;\ l = 1, \ldots, N \end{aligned} \tag{7.4.3}$$

When $\varepsilon = 0$, Eq. (7.4.3) is a general Hamiltonian system. If the rank of the system structure matrix is $r = 2n$, thus the general Hamiltonian system has $M = m - 2n$ Casimir functions ($C_1, \ldots, C_M$). If just $C_1, \ldots, C_M$ and H are integral invariants in the general Hamiltonian system, the general Hamiltonian system is completely non-integrable.

In a quasi non-integrable general Hamiltonian system, the Hamiltonian function H, and the M Casimir functions, are slowly varying processes. By applying the Itô differential rule, we derive an Itô stochastic differential equation for H and $C_1, \ldots, C_M$ from (7.4.3):

$$\begin{aligned} dC_s &= \varepsilon \left\{ d_{ij} \frac{\partial C_s}{\partial X_i} \frac{\partial H}{\partial X_j} + \sigma_{il}\sigma_{jl} \frac{\partial^2 C_s}{\partial X_i \partial X_j} \right\} dt + \varepsilon^{1/2} \sigma_{il} \frac{\partial C_s}{\partial X_i} dB_l(t) \\ dH &= \varepsilon \left\{ d_{ij} \frac{\partial H}{\partial X_i} \frac{\partial H}{\partial X_j} + \sigma_{il}\sigma_{jl} \frac{\partial^2 H}{\partial X_i \partial X_j} \right\} dt + \varepsilon^{1/2} \sigma_{il} \frac{\partial H}{\partial X_i} dB_l(t) \\ s &= 1, \ldots, M;\ i,j = 1, \ldots, m;\ l = 1, \ldots, N \end{aligned} \tag{7.4.4}$$

The variables $X_1, \ldots, X_{m-M-1}$ in the (7.4.4) are substituted by H and $C_1, \ldots, C_M$. According to the Khasminiskii limit theorem, the average Itô equation can be obtained by the deterministic average of the fast variable $X_1, \ldots, X_{m-M-1}$

$$\begin{aligned} dC_s &= \varepsilon U_s(C_1, \ldots, C_M, H) dt + \varepsilon^{1/2} \bar{\sigma}_{sl} dB_l(t) \\ dH &= \varepsilon U_H(C_1, \ldots, C_M, H) dt + \varepsilon^{1/2} \bar{\sigma}_{Hl} dB_l(t) \\ s &= 1, \ldots, M;\ l = 1, \ldots, N \end{aligned} \tag{7.4.5}$$

Wherein

$$
\begin{aligned}
U_s &= \frac{1}{T}\int\limits_{\Omega}\left[\left(d_{ij}\frac{\partial C_s}{\partial X_i}\frac{\partial H}{\partial X_j}+\sigma_{il}\sigma_{jl}\frac{\partial^2 C_s}{\partial X_i\partial X_j}\right)/[X_1,H]\right]\mathrm{d}X_1\ldots\mathrm{d}X_{m-M-1}\\
U_H &= \frac{1}{T}\int\limits_{\Omega}\left[\left(d_{ij}\frac{\partial H}{\partial X_i}\frac{\partial H}{\partial X_j}+\sigma_{il}\sigma_{jl}\frac{\partial^2 H}{\partial X_i\partial X_j}\right)/[X_1,H]\right]\mathrm{d}X_1\ldots\mathrm{d}X_{m-M-1}\\
b_{s_1s_2} &= \bar{\sigma}_{s_1l}\bar{\sigma}_{s_2l}=\frac{2}{T}\int\limits_{\Omega}\left[\sigma_{il}\frac{\partial C_{s_1}}{\partial X_i}\sigma_{jl}\frac{\partial C_{s_2}}{\partial X_j}/[X_1,H]\right]\mathrm{d}X_1\ldots\mathrm{d}X_{m-M-1}\\
b_{s_1H} &= \bar{\sigma}_{s_1l}\bar{\sigma}_{Hl}=\frac{2}{T}\int\limits_{\Omega}\left[\sigma_{il}\frac{\partial C_{s_1}}{\partial X_i}\sigma_{jl}\frac{\partial H}{\partial X_j}/[X_1,H]\right]\mathrm{d}X_1\ldots\mathrm{d}X_{m-M-1}\\
b_{HH} &= \bar{\sigma}_{Hl}\bar{\sigma}_{Hl}=\frac{2}{T}\int\limits_{\Omega}\left[\sigma_{il}\frac{\partial H}{\partial X_i}\sigma_{jl}\frac{\partial H}{\partial X_j}/[X_1,H]\right]\mathrm{d}X_1\ldots\mathrm{d}X_{m-M-1}\\
T &= \int\limits_{\Omega}\frac{\mathrm{d}X_1\ldots\mathrm{d}X_{m-M-1}}{[X_1,H]}\\
\Omega &= \{(X_1,\ldots,X_{m-M-1})|C_s(\mathbf{X})\le C_s, H(\mathbf{X})\le H\}
\end{aligned}
\tag{7.4.6}
$$

By the average Itô equation (7.4.5), we can obtain an average FPK equation as follows:

$$
\frac{\partial p}{\partial t}=\frac{\partial(U_s p)}{\partial C_s}-\frac{\partial(U_H p)}{\partial H}+\frac{1}{2}\frac{\partial^2(b_{s_1s_2}p)}{\partial C_{s_1}\partial C_{s_2}}+\frac{\partial^2(b_{s_1H}p)}{\partial C_{s_1}\partial H}+\frac{1}{2}\frac{\partial^2(b_{HH}p)}{\partial H^2}
\tag{7.4.7}
$$

Equation (7.4.7) is solved under certain initial conditions and boundary conditions.

7.4.2 Stochastic Optimal Control Strategy Based on Excitation Adjustment

Based on some approximate assumptions, the generator adopts a third-order model. Considering the controlled single-machine infinite bus system, the electromechanical transient process model under stochastic disturbance is[15]:

$$
\begin{cases}
\mathrm{d}\delta = \omega_N\omega\mathrm{d}t\\
\mathrm{d}\omega = \dfrac{1}{M}\left(P_m - D\omega - a_1E_q'\sin\delta + a_2\sin 2\delta + \xi(t)\right)\mathrm{d}t\\
\mathrm{d}E_q' = \dfrac{1}{T_{d0}'}\left(-bE_q' + c\cos\delta + u\right)\mathrm{d}t
\end{cases}
\tag{7.4.8}
$$

Wherein

$$a_1 = \frac{1}{x'_{d\Sigma}} V_s, a_2 = \frac{x_q - x'_d}{2x_{q\Sigma} x'_{d\Sigma}} V_s^2$$
$$b = \frac{x_{d\Sigma}}{x'_{d\Sigma}}, c = \frac{x_d - x'_d}{x'_{d\Sigma}} V_s$$
$$u = E_{fds} + u_f$$
$$x_{d\Sigma} = x_d + x_T + x_L$$
$$x'_{d\Sigma} = x'_d + x_T + x_L$$
$$x_{q\Sigma} = x_q + x_T + x_L$$

In the equation: ω is the deviation of the generator rotor angular speed and synchronous speed, δ is the power angle, M is the generator inertia time constant, P_{m} is the generator mechanical power, E'_q is the q-axis transient electric potential, T'_{d0} is the generator excitation winding time constant, E_{fds} is the generator excitation voltage at the steady state, V_{s} is infinite bus voltage, x_{d} is d-axis reactance, x_{q} is q-axis reactance, x'_d is d-axis transient reactance, x_{T} is transformer reactance, x_{L} is the reactance of the transmission line, and $\xi(t)$ is Gaussian white noise with a mean of 0 and a variance of σ^2; u_{f} is the system control variable.

In order to apply the stochastic averaging method, we need to convert the (7.4.8) into a Stratonovich stochastic differential equation, where $\xi(t)$ needs to be expressed by an independent standard Gaussian white noise $W(t)$, so that (7.4.8) can be expressed as:

$$\begin{cases} \mathrm{d}\delta = \omega_N \omega \mathrm{d}t \\ \mathrm{d}\omega = \dfrac{1}{M}\left(P_m - D\omega - a_1 E'_q \sin\delta + a_2 \sin 2\delta\right)\mathrm{d}t + \dfrac{\sigma}{M} W(t)\mathrm{d}t \\ \mathrm{d}E'_q = \dfrac{1}{T'_{d0}}\left(-bE'_q + c\cos\delta + u\right)\mathrm{d}t \end{cases} \tag{7.4.9}$$

According to the theory of stochastic dynamics, the Itô-type stochastic differential equation of (7.4.9) is:

$$\begin{cases} \mathrm{d}\delta = \omega_N \omega \mathrm{d}t \\ \mathrm{d}\omega = \dfrac{1}{M}\left(P_m - D\omega - a_1 E'_q \sin\delta + a_2 \sin 2\delta\right)\mathrm{d}t + \dfrac{\sigma}{M} \mathrm{d}B(t) \\ \mathrm{d}E'_q = \dfrac{1}{T'_{d0}}\left(-bE'_q + c\cos\delta + u\right)\mathrm{d}t \end{cases} \tag{7.4.10}$$

The Itô-type stochastic differential equation shown in (7.4.10) is a three-dimensional differential equation which is an odd-dimensional system. It is different from that the system dimension required by a quasi Hamiltonian system, which is an even number. Therefore, this is a quasi general Hamiltonian system.

The Hamiltonian function of the single-machine infinite bus system is as shown in (7.4.11).

$$\begin{aligned} H = & \frac{1}{2}M\omega_N\omega^2 + P_m(\delta_s - \delta) + a_1\left(E_{qs}^{'}\cos\delta_s - E_q^{'}\cos\delta\right) \\ & + \frac{1}{2}a_2(\cos 2\delta - \cos 2\delta_s) + \frac{a_1 b}{2c}\left(E_q^{'} - \frac{1}{b}E_{fds}\right)^2 \end{aligned} \quad (7.4.11)$$

In the equation: δ_s is the generator power angle at steady state, and $E_{qs}^{'}$ is the generator q-axis transient voltage at steady state.

The Hamiltonian function shown in (7.4.11) is composed of kinetic energy, potential energy, and magnetic potential energy, and is a representation of system energy.

A partial derivative of energy H for each state variable can be obtained based on (7.4.11):

$$\begin{cases} \dfrac{\partial H}{\partial \omega} = M\omega_N\omega \\ \dfrac{\partial H}{\partial \delta} = -P_m + a_1 E_q^{'}\sin\delta - a_2\sin 2\delta \\ \dfrac{\partial H}{\partial E_q^{'}} = -a_1\cos\delta + \dfrac{a_1 b}{c}\left(E_q^{'} - \dfrac{1}{b}E_{fds}\right) \end{cases} \quad (7.4.12)$$

Whereas $dE_q^{'} = 0$ at steady state, we can obtain that $bE_{qs}^{'} - c\cos\delta_s = E_{\text{fds}}$, and by substituting (7.4.12) into (7.4.10), we can get the quasi general Hamiltonian equation for the single-machine infinite bus system:

$$\begin{cases} \mathrm{d}\delta = \dfrac{1}{M}\dfrac{\partial H}{\partial \omega}\mathrm{d}t \\ \mathrm{d}\omega = \left(-\dfrac{1}{M}\dfrac{\partial H}{\partial \delta} - \dfrac{D}{M^2\omega_N}\dfrac{\partial H}{\partial \omega}\right)\mathrm{d}t + \dfrac{\sigma}{M}\mathrm{d}B(t) \\ \mathrm{d}E_q^{'} = \left(-\dfrac{c}{a_1 T_{d0}^{'}}\dfrac{\partial H}{\partial E_q^{'}} + \dfrac{1}{T_{d0}^{'}}u_f\right)\mathrm{d}t \end{cases} \quad (7.4.13)$$

For a general Hamiltonian system, the Hamiltonian function may not be a maximal or a minimal value at the equilibrium point. To analyze the stability of the system, an appropriate function is added to the Hamiltonian function so that the new Hamiltonian function at the equilibrium point takes a maximal or minimal value, and the new function can be applied instead of the original Hamiltonian function to determine the stability of the system. Such additional function is called the Casimir function[16].

The Hamiltonian function shown in (7.4.11) has a Casimir function as shown in (7.4.14).

$$C = \frac{a_1 b}{2c}\left(E_q^{'} - \frac{1}{b}E_{fds}\right)^2 \tag{7.4.14}$$

Obviously, this Casimir function is also a part of the system energy. According to (7.4.14), we can obtain:

$$\frac{\partial C}{\partial E_q^{'}} = \frac{a_1 b}{c}\left(E_q^{'} - \frac{1}{b}E_{fds}\right) \tag{7.4.15}$$

If there is no stochastic disturbance term $\sigma/M\mathrm{d}B(t)$ in (7.4.13), and according to the total differential equation, we can obtain:

$$\begin{cases} \mathrm{d}H = \dfrac{\partial H}{\partial \delta}\mathrm{d}\delta + \dfrac{\partial H}{\partial \omega}\mathrm{d}\omega + \dfrac{\partial H}{\partial E_q^{'}}\mathrm{d}E_q^{'} \\ \mathrm{d}C = \dfrac{\partial C}{\partial E_q^{'}}\mathrm{d}E_q^{'} \end{cases} \tag{7.4.16}$$

When considering the stochastic disturbance term, it is necessary to add the Wong-Zakai correction term to the first expression of (7.4.16) so as to obtain:

$$\begin{cases} \mathrm{d}H = \dfrac{\partial H}{\partial \delta}\dfrac{1}{M}\dfrac{\partial H}{\partial \omega}\mathrm{d}t + \dfrac{\partial H}{\partial \omega}\left(-\dfrac{1}{M}\dfrac{\partial H}{\partial \delta} - \dfrac{D}{M^2\omega_N}\dfrac{\partial H}{\partial \omega}\right)\mathrm{d}t \\ \quad + \dfrac{\partial H}{\partial \omega}\dfrac{\sigma}{M}\mathrm{d}B(t) + \dfrac{\partial H}{\partial E_q^{'}}\left(-\dfrac{c}{a_1 T_{d0}^{'}}\dfrac{\partial H}{\partial E_q^{'}} + \dfrac{1}{T_{d0}^{'}}u_f\right)\mathrm{d}t \\ \quad + \dfrac{1}{2}\dfrac{\sigma^2}{M^2}\dfrac{\partial^2 H}{\partial \omega^2}\mathrm{d}t \\ \mathrm{d}C = \dfrac{\partial C}{\partial E_q^{'}}\left(-\dfrac{c}{a_1 T_{d0}^{'}}\dfrac{\partial H}{\partial E_q^{'}} + \dfrac{1}{T_{d0}^{'}}u_f\right)\mathrm{d}t \end{cases} \tag{7.4.17}$$

Substituting (7.4.12) and (7.4.14) into (7.4.17), and eliminating the partial derivative on the right side of the equation, and adjusting it, we can obtain after arranging:

$$\begin{cases} \mathrm{d}H = \left(-D\omega_N\omega^2 + \dfrac{\sigma^2\omega_N}{2M}\right)\mathrm{d}t + \sigma\omega_N\omega\mathrm{d}B(t) \\ \qquad - \dfrac{c}{a_1 T'_{d0}}\left[-a_1\cos\delta + \dfrac{a_1 b}{c}\left(E'_q - \dfrac{1}{b}E_{fds}\right)\right]^2 \mathrm{d}t \\ \qquad + \dfrac{u_f}{T'_{d0}}\left[-a_1\cos\delta + \dfrac{a_1 b}{c}\left(E'_q - \dfrac{1}{b}E_{fds}\right)\right]\mathrm{d}t \\ \mathrm{d}C = -\dfrac{c}{a_1 T'_{d0}}(-a_1\cos\delta)\dfrac{a_1 b}{c}\left(E'_q - \dfrac{1}{b}E_{fds}\right)\mathrm{d}t \\ \qquad - \dfrac{c}{a_1 T'_{d0}}\left(\dfrac{a_1 b}{c}\left(E'_q - \dfrac{1}{b}E_{fds}\right)\right)^2 \mathrm{d}t \\ \qquad + \dfrac{u_f}{T'_{d0}}\dfrac{a_1 b}{c}\left(E'_q - \dfrac{1}{b}E_{fds}\right)\mathrm{d}t \end{cases} \tag{7.4.18}$$

According to (7.4.18), we can obtain the energy average Itô equation of the system as:

$$\begin{cases} \mathrm{d}H = \left(\bar{m}_H(H_\Omega, C_\Omega) + \left\langle \dfrac{u_f}{T'_{d0}}\dfrac{\partial H}{\partial E'_q}\right\rangle\right)\mathrm{d}t + \sigma_{HH}(H_\Omega, C_\Omega)\mathrm{d}B(t) \\ \mathrm{d}C = \left(\bar{m}_C(H_\Omega, C_\Omega) + \left\langle \dfrac{u_f}{T'_{d0}}\dfrac{\partial C}{\partial E'_q}\right\rangle\right)\mathrm{d}t \end{cases} \tag{7.4.19}$$

where:

$$\bar{m}_H(H_\Omega, C_\Omega) = \frac{1}{T(H_\Omega, C_\Omega)}\int_\Omega \left(\left(-D\omega_N\omega^2 + \frac{\sigma^2\omega_N}{2M}\right)\right.$$
$$\left. - \frac{c}{a_1 T'_{d0}}\cdot\left[-a_1\cos\delta + \frac{a_1 b}{c}\left(E'_q - \frac{1}{b}E_{fds}\right)\right]^2\right) \Big/ \frac{\partial H}{\partial \omega}\mathrm{d}\delta$$

$$\bar{m}_C(H_\Omega, C_\Omega) = \frac{1}{T(H_\Omega, C_\Omega)}\int_\Omega -\frac{c}{a_1 T'_{d0}}\left[-a_1\cos\delta - \frac{a_1 b}{c}\left(E'_q - \frac{1}{b}E_{fds}\right)\right]$$
$$\times \frac{a_1 b}{c}\left(E'_q - \frac{1}{b}E_{fds}\right) \Big/ \frac{\partial H}{\partial \omega}\mathrm{d}\delta$$

$$\sigma^2_{HH}(H_\Omega, C_\Omega) = \frac{1}{T(H_\Omega, C_\Omega)}\int_\Omega (\sigma\omega_N\omega)^2 \Big/ \frac{\partial H}{\partial \omega}\mathrm{d}\delta$$

$$\left\langle \frac{u_f}{T'_{d0}}\frac{\partial H}{\partial E'_q}\right\rangle = \frac{1}{T(H_\Omega, C_\Omega)}\int_\Omega \frac{u_f}{T'_{d0}}\frac{\partial H}{\partial E'_q} \Big/ \frac{\partial H}{\partial \omega}\mathrm{d}\delta$$

$$\left\langle \frac{u_f}{T_{d0}'} \frac{\partial C}{\partial E_q'} \right\rangle = \frac{1}{T(H_\Omega, C_\Omega)} \int\limits_{\Omega} \frac{u_f}{T_{d0}'} \frac{\partial C}{\partial E_q'} / \frac{\partial H}{\partial \omega} \mathrm{d}\delta$$

$$T(H_\Omega, C_\Omega) = \int\limits_{\Omega} 1 / \frac{\partial H}{\partial \omega} \mathrm{d}\delta$$

$$\Omega = \left\{ \delta | H\left(0, \delta, E_q'\right) < H_\Omega, C\left(0, \delta, E_q'\right) < C_\Omega \right\}$$

The integral region Ω is the bounded fluctuation region of the system energy. And since the control law u_{f} has not yet been determined, the control law in (7.4.19) can not be averaged for the time being, and further explanation will be given later.

As is shown in (7.4.20), the performance index is taken as the probability that the energy stays within the bounded fluctuation region, and the goal of the optimal control is to maximize the probability.

$$J\left(u_f\right) = P\left\{ H\left(s, u_f\right) \in [0, H_{\max}), C\left(s, u_f\right) \in [0, C_{\max}), t < s \leq t_f \right\} \tag{7.4.20}$$

where: $H_{\max}$ represents the maximum value of the system bounded fluctuation region energy, $C_{\max}$ represents the maximum value of Casimir function corresponding to the $H_{\max}$, u_{f} represents the control law, $u_{\mathrm{f}} \in U$ represents the control constraint, the form of which depends on the controller, and t represents the beginning time of the control, t_{f} represents the termination time of the control.

According to the reliability performance index shown in (7.4.20), the stochastic optimal control problem with the goal of maximum reliability is researched, and the value function is defined as:

$$\begin{aligned} V(H_\Omega, C_\Omega, t) = \sup_{u \in U} P\{ & H\left(s, u_f\right) \in [0, H_{\max}), C\left(s, u_f\right) \in [0, C_{\max}), \\ & t < s \leq t_f | H\left(t, u_f\right) \in [0, H_{\max}), C\left(t, u_f\right) \in [0, C_{\max}) \} \end{aligned} \tag{7.4.21}$$

According to (7.4.21), the corresponding dynamic programming equation can be established:

$$\begin{aligned} \frac{\partial V}{\partial t} = & -\sup_{u \in U} \left(\frac{1}{2} \sigma_{HH}^2 (H_\Omega, C_\Omega) \frac{\partial^2 V}{\partial H^2} + \left[\bar{m}_H (H_\Omega, C_\Omega) + \left\langle \frac{u_f}{T_{d0}'} \frac{\partial H}{\partial E_q'} \right\rangle \right] \frac{\partial V}{\partial H} \right. \\ & \left. + \left[\bar{m}_C (H_\Omega, C_\Omega) + \left\langle \frac{u_f}{T_{d0}'} \frac{\partial C}{\partial E_q'} \right\rangle \right] \frac{\partial V}{\partial C} \right) \end{aligned} \tag{7.4.22}$$

The boundary conditions are:

$$V(H_{\max}, C_{\Omega}, t) = 0 \text{ and } V(H_{\Omega}, C_{\max}, t) = 0$$

$$V(0, C_{\Omega}, t) = \text{limited and } V(H_{\Omega}, 0, t) = \text{limited}$$

The final value condition is:

$$V(H_{\Omega}, C_{\Omega}, t_f) = 1, H \in [0, H_{\max}), C \in [0, C_{\max})$$

If there is a suitable control law, which makes the right side of (7.4.22) take a maximal value, such control law is the optimal control law.

Taking the control constraint like $|u_f/T'_{d0}| \leq K$, and when $|u_f/T'_{d0}| = K$, and the positive or negative property of the value of u_f/T'_{d0} makes:

$$\frac{u_f}{T'_{d0}}\left(\frac{\partial H}{\partial E'_q}\frac{\partial V}{\partial H} + \frac{\partial C}{\partial E'_q}\frac{\partial V}{\partial C}\right)$$

(the expression above) positive, the right side of (7.4.22) takes a maximal value, so the optimal control law is:

$$\frac{u_f^*}{T'_{d0}} = K\operatorname{sgn}\left(\frac{\partial H}{\partial E'_q}\frac{\partial V}{\partial H} + \frac{\partial C}{\partial E'_q}\frac{\partial V}{\partial C}\right) \tag{7.4.23}$$

By substituting (7.4.23) into (7.4.19) to replace u_f, and completing the stochastic average, we can obtain the average Itô equation coefficient of the controlled system, which is as shown in the equation below.

$$\begin{cases} \mathrm{d}H = m_H(H_{\Omega}, C_{\Omega})\mathrm{d}t + \sigma_{HH}(H_{\Omega}, C_{\Omega})\mathrm{d}B(t) \\ \mathrm{d}C = m_C(H_{\Omega}, C_{\Omega})\mathrm{d}t \end{cases} \tag{7.4.24}$$

where

$$m_H(H_{\Omega}, C_{\Omega}) = \bar{m}_H(H_{\Omega}, C_{\Omega}) + \frac{1}{T(H_{\Omega}, C_{\Omega})}\int_{\Omega} \frac{u_f^*}{T'_{d0}}\frac{\partial H}{\partial E'_q} \Big/ \frac{\partial H}{\partial \omega}\mathrm{d}\delta$$

$$m_C(H_{\Omega}, C_{\Omega}) = \bar{m}_C(H_{\Omega}, C_{\Omega}) + \frac{1}{T(H_{\Omega}, C_{\Omega})}\int_{\Omega} \frac{u_f^*}{T'_{d0}}\frac{\partial C}{\partial E'_q} \Big/ \frac{\partial H}{\partial \omega}\mathrm{d}\delta$$

The conditional reliability function satisfies the backward Kolmogorov equation:

$$\frac{\partial R}{\partial t} = m_H(H_0, C_0)\frac{\partial R}{\partial H_0} + m_C(H_0, C_0)\frac{\partial R}{\partial C_0} + \frac{1}{2}\sigma_{HH}^2(H_0, C_0)\frac{\partial^2 R}{\partial H_0^2} \quad (7.4.25)$$

The boundary conditions are:

$$R(t|H_{\max}, C_0) = 0 \text{ and } R(t|H_0, C_{\max}) = 0$$

The initial condition is:

$$R(0|H_0, C_0) = 1$$

where, $m_H(H_0,C_0)$, $m_C(H_0,C_0)$ and $\sigma_{HH}^2(H_0,C_0)$ are obtained by taking $H_\Omega = H_0$ and $C_\Omega = C_0$ for $m_H(H_\Omega,C_\Omega)$, $m_C(H_\Omega,C_\Omega)$ and $\sigma_{HH}^2(H_\Omega,C_\Omega)$ in (7.4.19). H_0 and C_0 represent the initial values of energy function and the Casimir function.

After the conditional reliability function is introduced, the probability of the system within the bounded fluctuation region can be obtained by analytical methods.

7.4.3 Simulation and Analysis of Control Effect

The parameters for the single-machine infinite bus system are as follows [15]:

$P_m = 0.9$, $a_1 = 1.2835$, $a_2 = 0.4170$, $b = 2.9479$, $c = 1.9253$, $M = 7$ s, $T'_{d0} = 8$ s, $D = 4.6889$, $\sigma = 0.01$, $\delta_0 = 1.1999$, $E'_{q0} = 0.9879$, $f = 60$ Hz, $E_{fds} = 2.2144$.

Take the energy boundary of the bounded fluctuation region of the system to be $H_{\max} = 0.070$, and use the Casimir function to replace E'_q in the Hamiltonian function. The corresponding system critical trajectory is as shown in Fig. 7.5.

According to Fig. 7.5, the critical value of Casimir function is $C_{\max} = 0.1870$, and the largest integral area is $\delta \in [0.9347, 1.5555]$.

Taking the stable equilibrium point of the single-machine infinite bus system as the initial operating point, $H_0 = 0.0551$ and $C_0 = 0.0551$. The conditional reliability function of the system is as shown in Fig. 7.6.

The discrete points in Fig. 7.6 are Monte Carlo solutions derived from the motion equation (7.4.8), in which the step length is 0.01 for 5000 Monte Carlo simulations; the continuous curves are the analytical solutions obtained by calculation. According to Fig. 7.6:

(1) The analytical solution of the system fits well with the Monte Carlo solution;
(2) By adding control, the probability of system energy within the bounded fluctuation region increases, which improves the reliability of the system in the bounded fluctuation region;

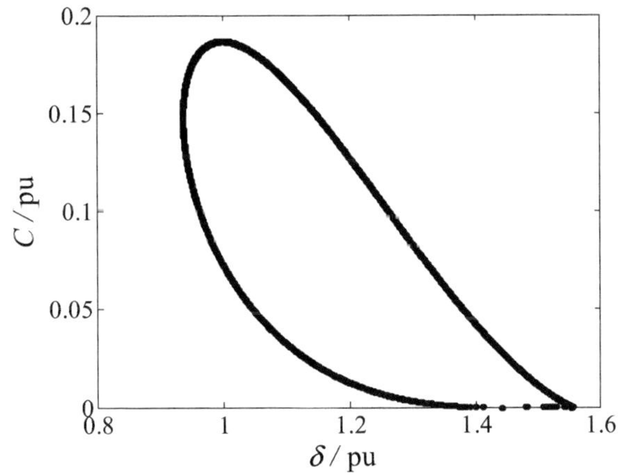

Fig. 7.5 Critical trajectory

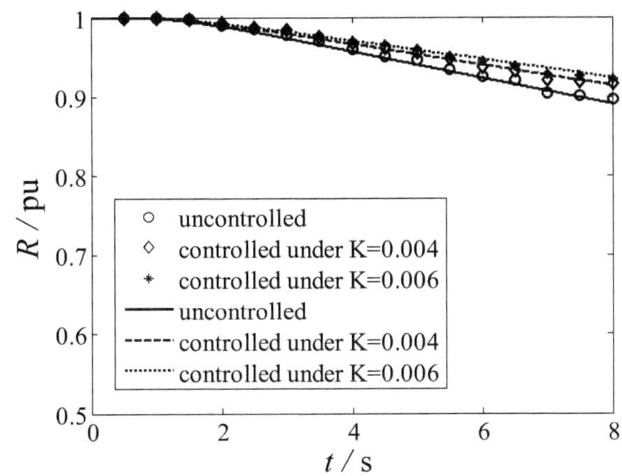

Fig. 7.6 Conditional reliability function

(3) At the same point, the greater the boundary K of the control constraint, the higher the probability of the system energy within the bounded fluctuation region.

Under different control constraint boundaries, the control effect and control efficiency of the control strategy calculated according to the power angle are shown in Fig. 7.7.

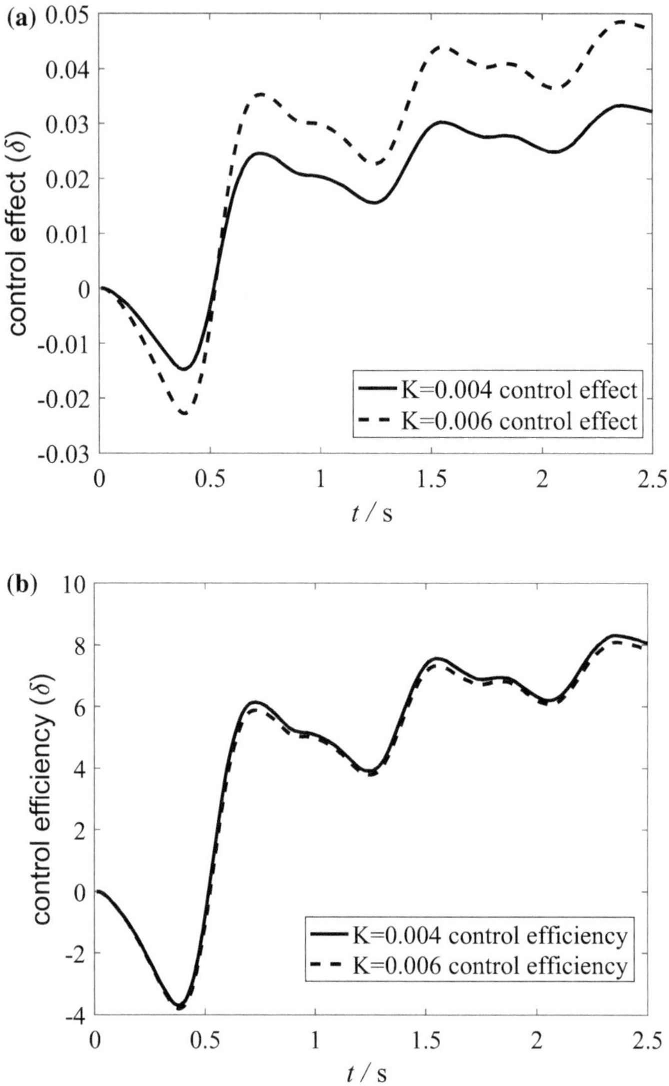

Fig. 7.7 **a** Control effect and **b** control efficiency

According to Fig. 7.7, the control effect is obviously increased with the expression of the control constraint boundary, and the control efficiency is slightly reduced, but basically consistent, which indicates that the control is effective. However, with respect to how to achieve balance between control effect and control efficiency, it still requires further study.

References

1. Chen, Lincong, and Weiqiu Zhu. 2010. First passage failure of dynamical power systems under random perturbations. *Science China(Technological Sciences)* 53 (9): 2495–2500.
2. Chen, Lincong, and Weiqiu Zhu. 2010. First passage failure of quasi non-integrable generalized Hamiltonian systems. *Archive of Applied Mechanics* 80 (8): 883–893.
3. Li, Hongyu, Ping Ju, Xinqi Chen, et al. 2015. Stochastic averaging method for quasi Hamiltonian system of multi-machine power systems. *Scientia Sinica(Technologica)* 45 (07): 766–772. (in Chinese).
4. Zhu, Weiqiu. 2003. *Nonlinear Stochastic Dynamics and Control: Hamilton Theoretical System Framework*. Beijing: Science Press. (in Chinese).
5. Zhou, Haiqiang, Ping Ju, Yusheng Xue, et al. 2016. Transient stability analysis of stochastic power system based on quasi-Hamiltonian system theory. *Automation of Electric Power Systems* 40 (19): 9–14. (in Chinese).
6. Ju, Ping, Hongyu Li, Yusheng Xue, et al. 2013. Stochastic models for study of electromechanical transient process in power systems. *Journal of Hohai University (Natural Sciences)* 41 (06): 536–541. (in Chinese).
7. Huang, Zhilong and Weiqiu Zhu. 2004. Stochastic averaging method for quasi general Hamilton system. *Proceedings of the Seventh National Academic Conference on Nonlinear Dynamics and the Ninth National Academic Conference on Nonlinear Vibration*, 399–403. (in Chinese).
8. Zhu, Weiqiu and Zuguang Ying. 2013. Advances in research on nonlinear stochastic optimal control of quasi-Hamiltonian systems. *Advances in Mechanics* 43 (1): 39–55. (in Chinese).
9. Liu, Yongfei, Ping Ju, Yusheng Xue, et al. 2014. Calculation analysis on power system characteristics under random excitations. *Automation of Electric Power Systems* 38 (9): 137–142. (in Chinese).
10. Li, Hongyu, Ping Ju, Yiping Yu, et al. 2015. Bounded fluctuations region and analytic method of intra-region probability in power system under stochastic excitations. *Proceedings of the CSEE* 35 (14): 3561–3568. (in Chinese).
11. Zhu, Weiqiu and Guoqiang Cai. 2017. *Introduction to Stochastic Dynamics*. Beijing: Science Press. (in Chinese).
12. Lu, Qiang, Shengwei Mei and Yuanzhang Sun. 2008. *Nonlinear Control of Power Systems*. Beijing: Tsinghua University Press. (in Chinese).
13. Ni, Yixin, Shousun Chen and Baolin Zhang. 2002. *Dynamical Analysis Theory for Power System*. Beijing: Tsinghua University Press. (in Chinese).
14. Kunder, Prabha. 1994. *Power System Stability and Control*. New York, NY, USA.
15. Sun, Yuanzhang, Xiaohong Jiao and Tielong Shen. 2007. *Nonlinear Robust Control of Power Systems*. Beijing: Tsinghua University Press. (in Chinese).
16. Li, Jibin, Xiaohua Zhao and Zhengrong Liu. 2007. *General Hamilton System Theory and Its Applications (Second Edition)*. Beijing: Science Press. (in Chinese).